Celluloid

Collectibles From the Dawn of Plastics

Robert Brenner

Schiffer Publishing Ltd®

4880 Lower Valley Road, Atglen, PA 19310 USA

Dedication

It is to Mary Lou Holt that this book is dedicated. Without her early appreciation and love for celluloid, many of the precious items illustrated in this book might have been lost forever.

Library of Congress Catalog Card Number: 98-83290

Book Design by Anne Davidsen
Type set in University Roman BD / Souvenir Lt BT

ISBN: 0-7643-0833-5
Printed in China
1 2 3 4

Published by Schiffer Publishing Ltd.
4880 Lower Valley Road
Atglen, PA 19310
Phone: (610) 593-1777
Fax: (610) 593-2002
E-mail: Schifferbk@aol.com
Please visit our web site catalog at
www.schifferbooks.com

This book may be purchased from the publisher.
Include $3.95 for shipping.
Please try your bookstore first.
We are interested in hearing from authors
with book ideas on related subjects.
You may write for a free catalog.

In Europe, Schiffer books are distributed by
Bushwood Books
6 Marksbury Rd.
Kew Gardens
Surrey TW9 4JF England
Phone: 44 (0)181 392-8585
Fax: 44 (0)181 392-9876
E-mail: Bushwd@aol.com

Contents

Acknowledgments

Without the generous understanding, motivation, and help of countless individuals this book would have never been possible. I especially wish to thank Doug Congdon-Martin for his photography expertise which is very evident. Hours upon hours of careful handling and delicate maneuvering were necessary to photograph the hundreds of celluloid items contained herein.

Equally important is Mary Lou Holt. The vast majority of the celluloid items photographed by Doug Congdon-Martin were hers. Mary Lou's interest in celluloid goes back to a period when precious few either knew anything about celluloid or cared to place it into their collections. Her insightful collecting can now be shared with all who peruse the pages of this book. Doug, my wife Sharon, and I are forever indebted to Mary Lou for allowing us to literally take over her house for almost a week while all these items were carefully documented and photographed.

Also adding to the collection of items photographed was Maxine Caldwell. Both Maxine and Mary Lou displayed true southern hospitality in so many ways.

Thanks also go out to Julie Robinson, a plastics historian, who helped perfect many of the details for this text.

Special thanks go to my wife Sharon, who spent hours upon hours helping with various aspects of this book as well. Sharon also picked up other household duties while I was writing. As usual, her editorial skills and suggestions were of tremendous help in this project.

Where would the world be without qualified and gifted librarians? Once again, a project of this magnitude would be impossible without the help of so many different librarians. Even though Princeton is rural and somewhat isolated geographically, our very qualified staff goes far in bridging this gap. Thanks go to Lorraine Cederholm, our competent head librarian, who spends countless hours helping everyone who asks for her assistance. Princeton's head librarian is far above her league in intelligence, dedication, and professional skills. Thanks also go to other staff and volunteers who helped fill the many requests required to complete this book: Shirley Hamaishi, Tammy Kiefer, Ruby Keller, and Marge Philbrick.

Thanks go to the following friends and associates who provided so much help in many different ways: Jim and Roberta Fiene, Laura and Craig Beane, Mike Makurat, Brian Swanson, and all who have touched our lives.

Finally, I wish to thank all those who have gone before us who have created these creative and colorful celluloid items for so many years. But most of all, we all need to thank those people who cared enough to save these very fragile celluloid works of art so that our generation might appreciate their beauty.

Chapter 1.
A Short History of Celluloid

Collectors of holiday memorabilia long ago discovered items made of celluloid. While many individuals found these objects somewhat garish and gaudy, there were those who found these whimsical and creative items worthy of collecting. In the early days, collectors labeled these items "early plastic" or "soft plastic," not fully aware of the term "celluloid." Enticed by their many forms—rabbits driving cars, Santas in sleighs, witches upon fat pumpkins, and even tiny dolls—collectors began to inspect these celluloid objects more carefully and soon came to appreciate their fragility and creativity. Thus an entirely new category of collecting came to the forefront as countless collectors began to amass collections of these celluloid items.

Not only holiday collectors have an interest in celluloid items; there are countless other collectors who find this material fascinating. Collectible celluloid pieces seem to fall into four principal categories:

toys: including holiday items, dolls, figures, and animals;

advertising premiums: including buttons, signs, and utilitarian advertising giveaways;

decorative items: including photograph albums, boxes, jewelry boxes, and lamps; and ...

fancy goods: including jewelry, hat pins, and ornamental hair combs.

While the scientific, definitive history of celluloid is left to scientists and those more versed in the history of this early plastic material, holiday enthusiasts and collectors can more fully appreciate these early marvels if a "light history" is presented. Thus it is in that vein that we proceed to partially explore the wonderful history of celluloid. While Alexander Parkes and John Wesley Hyatt are the primary names mentioned when one speaks of celluloid, history has recorded other names as well.

Alexander Parkes was born in 1813 in Birmingham, England. While growing up amidst ideology and invention, he was not specifically a chemistry and physics scholar. However, he was a metallurgist, craftsman, and inventor. This lack of scientific knowledge was not unusual at this period in English history. Parkes spent a great deal of time in the manufacturing of natural rubber just when great strides were being made with the discovery of vulcanization. This stirred a curiosity in Parkes to explore other substances which could give similar results to those of rubber. By studying cellulose nitrate, obtained in 1845 in Basle by C.F. Schoenbine, Parkes developed a new material which could be used in a solid, plastic or fluid state, which was at times rigid like ivory, opaque, flexible, water resistant, colorable, and could be used for utensils and tools like metals.

John Wesley Hyatt from Starkey, in the United States, soon followed in the steps of Parkes. Hyatt was also an inventor as well as a printer by profession. Reading a Phelan and Collander poster in Albany, New York, announcing a prize of $10,000 for anyone capable of producing a new material which could replace the ivory used for billiard balls, Hyatt threw himself head-first into the search for "artificial ivory" or any new material which could meet the industrial demands. In 1869, he was successful with a compound employing a cellulose nitrate base, just as Parkes had been before him, only on the opposite side of the Atlantic.

Celluloid is the "generic" term applied to this early plastic. Where does the term "celluloid" come from? Two different stories have evolved over the years. One story is that the word is a contraction for "colloid of cellulose" and the other story is that it is a combination of the words "cellulose and oid" (a Greek word for "like").

Cellulose is a large molecule consisting of many identical units called monomers. In cellulose, the monomer is cellobiose composed of two "simple" sugar molecules. When cellulose is treated with Nitric Acid, a chemical change takes place in the molecules. The cellulose molecule is "nitrated" and a molecular reconstruction called Polymerization takes place. The individual molecules link up into a chain-type structure. Hyatt was not a scientist and he never really understood the impact his discovery made on Polymer science.

Various trade names were used over the year when marketing celluloid items for sale. Some of those names include Xylonite (English term), Zylonite (American), Viscoloid, Ivorine, Coraline, Pulveroid, and Fiberloid. Others include Pyralin, Cellonite, and Pasbosene. All of these trade names were used in trade catalogs such as Butler Brothers, Montgomery Ward, and Sears when illustrating celluloid items in their catalog pages.

One of the first experimenters of celluloid was Dr. Pierson of New Orleans, who, among others, was seeking an imitation of horn, tortoise shell, and ivory. These natural materials were in short supply and scientists were attempting to find artificial replacements for them. In 1848,

Dr. Pierson discovered that a complex substance consisting chiefly of nitrocellulose and camphor could replace these scarce materials, but no industrial application was made until the 1860s.

Another early experimenter was the Swiss chemist Christian Shoenbine who had developed cellulose nitrate, obtained when cotton or wood fibers interacted with nitrate and sulfuric acids. The resulting moldable mass—which was highly flammable and was used primarily in the explosives industry—began to interest those who felt that the doughy substance might be used for a variety of purposes. However, Shoenbine's experiments resulted in little advancement, at that time, toward the development of plastic as we know it.

Alexander Parkes, an Englishman born on December 29, 1813, in Birmingham, Warwickshire, England, and who would die on June 29, 1890 outside of London, advanced plastics history remarkably in 1855 when, through experimentation, he created a form of cellulose nitrate which he termed "Parkesine." Parkesine was a compound of cellulose nitrate, naftalene (castor oil was essential as a plastizer), and camphor, which acted as thinner. Parkes was keenly interested in finding a material, which could replace all previously known natural substances such as horn, gutta percha, ivory, and tortoiseshell. He used this new material to make books covers, bracelets, buttons, combs, handles, and even pens ... among other things. Parkes took out 20 patents for moldable pyroxylin materials alone.

Parkes presented the first Parkesine items at the Great London Exhibition in 1861. However, he was unsuccessful in commercially producing items made from Parkesine.

Interestingly enough, his lack of success was due to his use of inferior raw materials which resulted in poor products. Parkes' company went out of business in 1868. Parkes gave Daniel Spill, another Englishman, his permission to produce the material he had invented.

Spill produced items under the names "Xylonite" and "Ivoride." He was in business from 1865 to 1891, but financing and other setbacks, with patent infringement against John Wesley Hyatt in American courts, brought about major reversals in his endeavors. Hyatt had accused Parkes of stealing his formula. Daniel Spill lost the infringement case, leaving Spill's celluloid manufacturing business in ruins.

Julie Robinson, a noted modern-day plastics historian, documents some of the early history of celluloid. She recounts the reasons for its perfection in the United States as follows. In the United States, celluloid's history quickly took a turn for the better. Billiard balls and elephants had a lot to do with this success. Around the 1860s, big-game hunters were killing so many elephants that makers of billiard balls began to worry about the extinction of the species and the loss of their ivory supply. Over 12,000 elephants were being slaughtered each year, and herds were decreasing at a rapid rate. It should be noted that it was a dangerous occupation as well. For every two elephants killed, it was reported that one hunter lost his life in the hunt. If elephants were to disappear, where would the industry obtain its much-needed ivory used in the manufacturing of billiard balls? English billiards master Michael Phelan was one person quite concerned about the rising cost of elephant ivory used in manufacturing billiard balls.

In 1863, Phelan and Collander, America's largest billiard supplier, offered a $10,000 prize to anyone who could create a substitute material with which to manufacture these balls. John Wesley Hyatt, an ambitious 26 year old printer from Albany, New York, discovered the advertisement and decided he could meet this challenge.

Hyatt worked in his print shop by day and experimented in the evenings and on weekends. He fervently hoped to be the first person to successfully manufacture this ivory substitute, and, of course, win the prize. His first experiments included combining common substances such as sawdust, paper flock, shellac, and varnish, but none of the compositions he created were hard enough to produce billiard balls.

But, in 1868, Hyatt made nitrocellulose by dissolving cotton in a mixture of nitrate and sulfuric acids. Then Hyatt added camphor and coloring matter, and molded the mixture in a die under heat and pressure. The Celluloid patent (actually a patent listed as "Improvement for Molding Solid Collodion) was registered in the United States in 1870. With that, Hyatt began manufacturing composition core pool balls made of paper-pulp and gum-shellac at the small factory he called The Hyatt Billiard Ball

Company. This composite material had molding potential, which helped propel Hyatt's business sucessfully; however, celluloid as we now know it was still somewhere over the horizon.

With his younger brother Charles, Hyatt founded a small checker manufacturing business employing his newly discovered material. The company grew at a steady pace and, in time, it was divided into two divisions: The Hyatt Manufacturing Company, which relocated to a factory on Beaker Street where John acted as the head of the company, and The Embossing Company, located at 4 & 6 Pruyn Street, with Charles acting as president.

The Embossing Company enjoyed huge successes, primarily due to Edward C. Schoonmaker, a charismatic salesman who became the driving force behind the company. The Embossing Company produced colorful, educational games that are highly collectible today. They included Dominoes, checkers, block puzzles, and Color Cubes. Most of these were produced between 1920 and 1940.

John Hyatt remained with his printing job and continually experimented with ivory substitutes as he *still* had not won the $10,000 prize. Throughout the years 1868 and 1869, Hyatt constantly continued his experimentation with other substances in the earnest hope of improving the Hyatt Billiard Ball. For he still had yet to discover the ivory substitute necessary to win the prize offered by Phelan and Collander. His discovery of celluloid was just around the corner. Actually Hyatt's discovery of celluloid came quite by accident. In that time period printers used a bottled solution of pyroxylin (nitrocellulose and alcohol) called "Collodion," that when brushed onto the printer's hands, dried into an elastic waterproof film. Thus paper cut and ink stained hands were avoided. While preparing for work one morning, Hyatt discovered that his bottle of Collodion had tipped and spilled some time in the evening before.

The resulting dried hunk of solid pyroxylin, about the size and thickness of his thumbnail was hard, yet pliable and resembled the density of the ivory he had been attempting to duplicate. It was then that he was extremely close to the success.

John relocated the Hyatt Billiard Ball Company to Grand and Plain streets in Albany, New York, and changed the name to The Albany Billiard Ball Company. There were some setbacks. The resulting billiard balls were both flammable and noisy, making a loud sound when the balls hit each other on the pool table. In fact, there are records of innkeepers who complained about this resulting noise, which sounded like a loud "bang." In fact, some balls exploded when struck with sufficient force. The sounds of those exploding billiard balls upset many a saloon owner and patron. In addition, there were other serious hazards with this nitrocellulose-based coating. Since it was highly flammable, close contact with a lit cigar caused these balls to burst into flames, which resulted in serious saloon fires. A dull film, which attracted dirt like a magnet, quickly formed on the billiard balls as well. Thus they were very difficult to keep clean as well as shiny. In spite of all their drawbacks, Hyatt's billiard balls were still much less expensive when compared with ivory balls. They were also much more plentiful. Thus billiards, once played only in the homes of the wealthy, now could be enjoyed by the general population.

John and his brother Isaiah, who coined the trade name "celluloid," created a substance which was, in essence, the same as the Parkesine and Xylonite materials previously created. Hyatt's improved formula helped him to win a long patent infringement case brought against him by John Spill. Spill had taken out American patents for celluloid in 1869 and 1870. According to Julie Robinson, while Parkes had given Spill the right to produce Parkesine, Spill then claimed the formula was his own and sold licensing rights to it in America under the name Xylonite. He went to battle in court with Hyatt. But Hyatt eventually won his case in 1884 and Spill lost his entire enterprise.

Hyatt's improved formula added camphor as a solvent and applied heat and pressure to the mixture to form his new plastic material from solid Collodion. Producing the raw material was a complex process. It included nitrating the cellulose to make pyroxylin, then adding nitrocellulose with camphor and other solvents, and then forming it into sheets, rods, and tubes. The first step was the most critical. If too much nitric acid was added, guncotton, a highly explosive material would result.

Celluloid is made from a very pure form of cellulose, obtained from cotton linters or unsized filter paper (tissue paper). The cellulose is subsequently converted in nitrocellulose, using Hyatt's method. Sulfuric acid was essential as it caused the cotton fiber to swell and absorb the nitric acid. Once the cellulose was nitrated, it was washed in alcohol and water; then mixed with camphor. This process varied somewhat from factory to factory. The most common process consisted of the raw material—raw or spun cotton, or paper cut or chopped into strips—being immersed in nitric acid for fifteen minutes to two hours, depending upon the character of the fibers and the temperature of the bath.

The cotton or paper, once converted into nitrocellulose, was taken out, wrung, and pressed to remove most of the adhering liquid. It was then restored to its original strength by the addition of concentrated nitric acid.

The nitrocellulose was washed in water and ground in a paper mill, then pressed as it passed between a plate and two rotating cylinders moving at a speed of 160 revolutions per minute. Success was on the horizon for Hyatt.

As a result, Hyatt and some associates formed the Celluloid Manufacturing Company in 1872, creating a machine for celluloid manufacturing. The company set up shop in an old cradle factory. The owners also commenced the building of a large complex of brick buildings in the vicinity of Broad and Ferry Streets, adjacent to the former cradle factory. Later, this area developed into the heart of Newark's industrial district. According to Julie Robinson, between 1872 and 1880, 23 different companies were granted production licenses from the Celluloid Mfg. Co. to manufacture finished articles using plastic materials supplied by the parent company. Hyatt worked closely with Charles Burroughs, whose name is quickly associated with the office machine business. In 1878, working in a harmonious relationship, Burroughs and Hyatt developed the first extrusion and compression molding machines. First this new material was used chiefly for dental plates. Unfortunately the dentures tasted like camphor and warped when exposed to hot liquids. Drink tea and the dentures curled. Then someone conceived of the idea of producing celluloid collars. Industry specialists had been attempting to waterproof collars since the 1860s. Not only could these collars be cleaned with a damp cloth, but also, according to early advertisements, the camphor content of celluloid had valuable medicinal qualities.

> "Testimonials are being continually offered from parties...who have suffered from many throat and lung diseases extant, that (the use of celluloid collars) has effected a permanent cure of these deadly ills and that they would not dare wear any other kind."

The first step in the process included forming blocks of celluloid from which thin sheets were planed. These resulting sheets were subsequently stamped to form trays, picture frames, or even cover wooden handles (to imitate the natural material Ivory). The French were most famous

for the "French Ivory" wares which most often were vanity sets and jewelry items. Their cream colored effect and delicate striations still cause some experienced collectors to think of these as true ivory. What most attracted consumers to these "French Ivory" items were the advertising claims that celluloid would not crack or discolor and was not affected by hot water temperatures. Most attractive to housewives was the inexpensive price when compared to more expensive materials.

In 1888, at the request of John Carbutt, a Philadelphia maker of photographic plates, Hyatt produced very thin, uniform sheets of celluloid. These replaced the heavy and fragile glass plates then in use.

Hyatt also was the first in the industry to employ the blow-molding process. In this process hollow tubes of celluloid, held in a hollow press were heated and then inflated with hot air until the tube transformed itself into the shape of the mold being employed. The result was a three-dimensional hollow object.

By the 1890s, fashion accessories in many forms were being produced. Fancy ornamental boxes, manicure and grooming sets, ornamental combs, and jewelry soon became popular gift items, especially to the middle class who enjoyed both the luxurious look and affordable pricing. By 1900, almost every home in the United States used celluloid in one form or another. Advertising premiums, eyeglasses, toys, and novelty items abounded.

Due to its almost instant popularity, celluloid soon began to be produced in Germany, France, Japan, and the United States by many different companies. In fact, much of the manufacturing was occurring at the very same time. Then, as now, different countries always vied to be the most innovative in producing items for consumers.

In France, celluloid was on the minds of many. Since the French produced many porcelain dolls, they were eager to experiment with less expensive materials from which to produce dolls and dolls' heads. In the mid-1870s, the Compagnie Franco-Americaine was established in Stains, just outside Paris. Later operating under Hyatt's American patents, it became the Compagnie Francaise du Celluloid. In 1887, Valmore Boitel obtained a French patent to make celluloid dolls. While the bodies were stuffed, their heads were manufactured of celluloid. Among other French companies was Petitcolin of Paris, which established factories later at Lilas, Seine, Etain, and Meuse. Dolls dressed in regional costumes were a specialty of the French. Thus many of these unmarked dolls could be of French manufacturing. One of the largest celluloid producers in France was Societe Industrielle du Celluloid.

Steeped in the production of toys and fancy goods for the American market, Germany was also deeply involved in early experiments with celluloid production. As early as 1863, celluloid appeared as a synthetic material in the manufacturing of dolls. In 1878, Magnus and Company of Berlin commenced the manufacturing of celluloid. Later that year, another plant was put into production at Offenbach, near Frankfurt.

While these factories did meet with financial success, the first successful factory was established in 1873 by the Rheinischen Harggummiwaarenfabric (previously, the Rheinische Gummi-und Celluloid Fabrik) at Mannheim. Fritz Jander oversaw the production of celluloid, but a disastrous fire in 1880 destroyed the main factory. The factory was quickly rebuilt and restored to full capacity operations. Their trademarks were the turtle and turtle in a diamond. They registered their turtle trademarks in France and Great Britain in 1914. In 1915, they registered this same mark in the United States. Rheinische Gummi-Celluloid also produced many different celluloid toys including the celluloid roly poly toys sold as penny toys in the United States. Dr. Paul Hunaeus manufactured roly poly penny toys for the company near Hanover, Germany. As early as 1895, the Lenel Bensinger Company of Mannheim was stamping its products with the familiar "turtle" mark as well—a mark they adopted as their own.

Most intriguing was the process perfected by the early 1920s which created inexpensive lightweight toys of celluloid. Injection molding was borrowed from Germany.

Injection molding was long employed for metal toys, but was not applied to the celluloid industry until 1919 when A. Eichergrun introduced a cellulose acetate molding powder and Hermann Buchholz invented an injection press to mold small pieces with a minute amount of material.

Air was blown between softened sheets of celluloid. The process, brought to perfection by Germany, was also employed in the United States and Japan. Two very thin sheets of celluloid were placed between two heated dies and welded together at the edges by pressure, forming an airtight unit. Forced air was then introduced between the heat-softened sheets. This air pushed them outward to line the cavity of the two-part mold. The result was a hollow one-piece, three-dimensional toy, human figure, animal, or doll. The toys ranged in size from as small as one-inch in height up to twelve-inches. The celluloid easily molded itself to the shape, detail, and texture of the mold into which it was blown. Depending upon the size of the toy, a single mold could contain from six to more than a hundred cavities. Thus many toys could be very efficiently produced. Less human handling in this process was a further incentive since the resulting toys could be sold for pennies. The Viscoloid Company of Leominster, Massachusetts, was America's leader in the manufacturing of these small blow-molded toys. They were primarily produced from about 1915 to the late 1920s.

Often times, though, several thin blown-molded parts were produced. These were necessary for the jointed pieces as well as for many of the toys which required multiple parts. Many times, in Europe, these parts were assembled by individual families or smaller factories which put the toys together by hand and subsequently painted the pieces. This is quite similar to the manufacturing process so often employed for so many different Christmas tree ornaments as well.

The United States was quick to respond to German ingenuity. By 1927, the Celluloid Corporation imported several Buchholz machines manufactured by Eckert & Ziegler in Germany. In 1927 the Celluloid Company listed over 3000 uses for its materials, which indicates its wide acceptance. It was in that year that the Celluloid Company developed cellulose acetate, a nonflammable molding powder that looked exactly like Celluloid when fabricated. This workable cellulose acetate molding powder worked well in compression molding with over 40 pounds of steam pressure.

At about this same time, unaware of what was occurring in the industry, Samuel Foster, Jr. of Foster Grant, a very small celluloid company in Leominster went to New York to examine some products produced in Germany using the injection mold process. Impressed with what he examined, he ordered ten of the Eckert & Ziegler ma-

chines. However, he canceled his order after two machines arrived and he quickly discovered that cellulose acetate did not work well with these machines. The Germans had used polystyrene which was much more expensive to use as a raw material. After years of modifying the machines, Foster was able to begin commercial injection molding of cellulose acetate in 1934.

Principal American competitors in the plastics industry included Fiberloid Company, Foster Grant Company, Arlington/Dupont Company, and Viscoloid Company. Trade names such as "Pyralin," "Fiberloid," and "Viscoloid" became a part of everyone's vocabulary. In 1910, the E. I. DuPont de Nemours Company entered the market with the purchase of Arlington Company.

The American Zylonite factory was built between Adams and North Adams, Massachusetts, in the late 1870s. American Zylonite is thought to have produced many of the earlier ivory-colored baby rattles and tiny dolls, even though the company was in existence for only six years.

In a 1914 issue of The *Journal of Industrial and Engineering Chemistry,* L. H. Baekeland, Ph. D., praised John Wesley Hyatt's innovative work with celluloid and his subsequent award of the Perkin Medal. The Perkin Medal was awarded by the Society of Chemical Industry to the 77 year old Hyatt for his inventions of Celluloid and a Sugar Cane Mill.

It was Baekeland himself who discovered, in 1907, how to control the union of phenol, or carbolic acid, and formaldehyde to make the synthetic resin known as Bakelite. As Baekeland continued, he states that Hyatt had discovered the important fact, on which the entire celluloid industry was based, that a mixture of nitrocellulose, camphor, and a small amount of alcohol, when properly prepared, became thermoplastic. Simply stated, the mixture becomes soft when treated, can be molded in a hydraulic press, and after cooling to ordinary atmospheric pressure it becomes hard and strong once again.

Hyatt did endure some threats from competitors and those who were jealous of his development of celluloid. His patents were assailed on the ground that others before him had used solvents and camphor in conjunction with nitrocellulose, but it was studiously omitted that his predecessors had used these ingredients under entirely different conditions for entirely different purposes, and could not produce in that way Hyatt's valuable technical effect. Hyatt was quite unfamiliar with chemistry. His knowledge of nitrocellulose was obtained in stages by his own experimenting.

On May 20, 1920, Hyatt died of a heart failure at his home in Short Hills, New Jersey. He was 82 years old and left behind two sons, Charles and Ralph, as well as his wife, Anna Taft Hyatt. He had a remarkable life. By the time of this death, he had over 238 patents issued in his name, 77 of those directly relating to the manufacturing of Celluloid.

Many different American companies were involved in the Celluloid industry. The Arlington Co., the makers of Pyralin, was owned by DuPont. Fiberoid Corporation also produced large amounts of celluloid and was eventually absorbed by Monsanto. Viscoloid of Leominster, Massachusetts (absorbed in time by Dupont), and the Celluloid Corporation of Newark, New Jersey (absorbed by Cellanese), also produced tons of the material every year.

The most important American manufacturer of celluloid after the original Hyatt Company was Merchant's Manufacturing Company which produced Pasbosene and Celonite. Located in Arlington, New Jersey, Merchant's actually was a combination of several small companies started in 1881. They built a plant in Arlington, New Jersey, in 1885, and the company was renamed the Arlington Manufacturing Company (later simplified to the Arlington Company). They renamed their product as "Pyralin."

The Celluloid Company was instrumental in product development in its years of operation. Between 1869 and 1891, over 61 different patents relating to the manufacturing of celluloid were granted to this company. The Celluloid Company was the result of a merger between the Celluloid Manufacturing Company, the Celluloid Brush Company, the American Zylonite Company, the Celluloid

Novelty Company, and some other minute companies. The Arlington Company was then its principal competitor. Two of the companies licensed by the Celluloid Manufacturing Company (CMC) to manufacture items were somewhat widely focused on producing different items for the market place. The Celluloid Novelty Company and the Celluloid Fancy Goods Company produced a wide variety of goods ranging from tape measures to soap dishes, to dice boxes, to various types of jewelry. In 1890, they attempted to absorb its most viable competitors.

Initially, Viscoloid Corporation was established in 1900 by Bernard W. Doyle, a Leominster, Massachusetts, entrepreneur, to manufacture celluloid for the various comb manufacturers located in the region. Doyle was a buyer of horn and tortoise shell (natural plastics) and had become familiar with pyroxylin plastic in France while on a buying trip. Viscoloid was later acquired by DuPont, which had also acquired the Arlington Company in 1915, then the main rival of Hyatt's Celluloid Company.

Fiberloid Company was first located in Newburyport, and later, Springfield, Massachusetts. DuPont Viscoloid purchased over a dozen injection molding machines in the early 1930s from Germany. Injection molding was at its acme in the mid-1930s when the Index Machinery Corporation of Cincinnati imported its first fully automatic press. Americans first witnessed such an automatic press at the Leipzig Fair of 1934. American molders such as Viscoloid soon realized its superiority. This machine was able to mold in fifteen seconds something which originally could take up to fifteen minutes to mold. Before 1939, over fifty such machines were imported into the United States. In the late 1930s, Monsanto absorbed Fiberloid.

When Japan waged a price competition war with Germany in the 1920s in celluloid manufacturing, Germany gradually phased itself out of celluloid sales. Japan then captured the largest share of this market as they flooded our department stores and mail order catalogs with endless assortments of toys each year. Part of Japan's success in this field was due to the fact that Japan held the world's largest natural resource for camphor and had a monopoly on the commodity.

Even though "Lucite" is a recognizable term today for clear acrylic objects, it originally referred to women's dresser sets or boudoir accessories. Lucite was merely one of a series of names given to various designs by the DuPont Company for these products. In 1929, Lucite was known for eight different boudoir sets named Crystal, Ming, Orchis, Wedgewood, Empire, Venetia, Diane, and Wedgwood. While the designs were given these names, all of these sets were made from "Lucite," pyroxylin plastic.

One of the largest producers of celluloid was the Irwin Toy Corporation of Leominster, Massachusetts. Irwin Toy Corporation was founded in the 1920s when Irwin Cohn purchased the old Viscoloid toy molds and used acetate and polystyrene to produce his toys. His company was called Great American Plastics and he placed his trademark over the old trademarks originally used by Viscoloid. Celluloid rattles and pinwheels were their first endeavors.

Celluloid became so very popular that ladies magazines as well as hobby and craft publications began to carry articles illustrating and explaining how anyone in their own home could use celluloid to fashion some very interesting items. Since celluloid was manufactured in so many colors and various weights (sheets), it caught on rapidly. The physical nature of celluloid was such that it could be carved, sawed, cut, turned, or manipulated in most any manner with ease. It did not split or crack readily, yet its texture was such that it was easily worked. Celluloid was also inexpensive.

Even trade journals sent to teachers included directions for producing items from celluloid. In a 1924 issue of *Industrial-Arts Magazine,* celluloid was referred to as "Ivory." Handbags with ivory tops, boxes made of ivory, bag covers, and combs were offered as suggestions with explicit directions so the objects could be fashioned by students in their manual arts classes.

A *Popular Mechanics* article in 1931 revealed that immersing celluloid in water at about 110 to 125 degrees for a few minutes would render the celluloid sheet quite flexible. Subsequently it could be bent; when cooled, it kept its shape. Other suggestions for celluloid craft work in that issue included napkin rings, a letter file, ruler, paper knife, pendant, earrings, and even a lemon fork. Subsequent articles related the fact that attractive handles for commonly used articles such as nail files, combs, pocketknives, and razors could be made by cementing together pieces of sheet celluloid.

Coloring the celluloid was also quite easy according to Kenneth Murray in *Popular Science Monthly*. First, the surface had to be thoroughly cleaned as a trace of oil, or even finger prints, could spoil the result. For a black color, the celluloid was soaked in a saturated solution of silver nitrate for an hour and then dried in bright sunshine. For red, the celluloid was cleaned with a diluted nitric acid, then soaked in a solution of aniline red dye which was made alkaline with the addition of an ounce of ammonia to each five ounces of solution. Other aniline dyes could be used in the same manner. A strong potassium bichromate solution would color the celluloid yellow or orange. Brown could be obtained with a potassium permanganate solution, with a pinch of baking soda added to each ounce to increase the color. Once dyed, the pieces were varnished. Once varnished, the surface would become practically fireproof.

In 1935, *Popular Science Monthly* provided directions for creating costume jewelry from sheet celluloid. They first suggested the celluloid be shaped with a hand scroll saw. When each piece of jewelry was finished, the creator would finish any rough edges with a file and then use fine sandpaper to obtain a smooth, satin appearance. The original gloss was easily restored with acetone. It was suggested that, while the acetone was wet, a little bronze powder of any color could be lightly dusted on the jewelry for extra effect. Imitation cameos of a cream-colored celluloid and earrings made from jade celluloid were actually quite attractive. as pictured in the magazine.

In 1926, Professor Ebner, Ph.D., discussed the riddle of cellulose. As he stated, "cellulose is the chief structural material of all plants, but also was quickly becoming the raw material of so many industries in the 1920s. It was difficult to scientifically investigate its physical and chemical constitution since it was insoluble in water or other solvents, and it was extremely sensitive to the action of acids and saline solutions."

Over the years, many other products were devised from celluloid. These included combs (1874), eyeglass frames (1878), toothbrush handles (1880), toiletware (1881), buttons (1880s), and some 25,000 products by the year 1936. Its use as photographic film was most significant in that it revolutionized the photography industry and most certainly was the impetus behind the movie industry. Because celluloid film came into contact with hot lights, it caused some disastrous fires.

In 1900, the Swiss scientists Henri and Camille Dreyfuss preferred a process in which acetic acid (vinegar) was substituted for the volatile nitric acid. They were among the first to perfect a cellulose acetate yarn in 1925 which was introduced in the United States by the Celanese Cooperation. The Celanese Corporation later became part of the Hoechst-Celanese Cooperation. This company absorbed the Celluloid Company in 1927. Their experiments helped pave the way for "safety film," even though it was not entirely flame proof. "Cellulose nitrate" now was replaced by "Cellulose acetate." Among the advantages of this new substance was its ability to be injected molded, a mechanical process which almost instantaneously produces many identical objects.

Phillip H. Smith mentioned celluloid in article in *Scientific American* in 1935. Smith states that no description of plastics would be complete without mention of nitro-cellulose and cellulose acetate. He goes on to state "cellulose-acetate is much like pyroxylin, but it does not burn explosively and therefore, sometimes substitutes for it. Its high dielectric properties and low water absorption make it an excellent insulator." He mentions the fact that cellulose acetate was best known commercially as Masuron, Tenite, and Lumarith.

By 1933, celluloid production was close to 12,000,000 pounds worth of materials. Over 200 colors were available, although pastels and ivory mostly prevailed as popular choices for items.

Celluloid combs were actually quite dangerous. In 1938, Dr. Howard Fox, a well-known dermatologist of New York City, warned women of the danger from combs used for waves. Most accidents occurred in the home when women used electric lamps, gas heaters, or stoves to dry their hair.

Since celluloid was very flammable, Lumarith was introduced in 1927. Made of cellulose acetate instead of nitrocellulose, this material was relatively slow burning. It could be molded more easily than celluloid and came in a wider variety of colors. Lumarith substituted for Celluloid in many applications including buttons, toys, jewelry, and photographic film. New plastics were also listed by *Newsweek* magazine in a June 6, 1936 article. These plastics included Bakelite, Durez, and Beetleware. Since these had lower production costs, their sales skyrocketed.

Cellulose acetate resulted from a long search for a nonflammable substance. Since cellulose nitrate was not safe to manufacture, store, or fabricate; companies had been searching for years for a suitable replacement. Like its predecessor, cellulose acetate was tough, flexible, and adapted to a wide variety of colorations, and molded easily when heated. Quickly it became the most advanced plastic of the 1930s. Nonflammability was its chief virtue. Its popularity in molding inexpensive toys was quickly discovered since it melted without any danger of fire.

Backing up a bit, after World War I, the Celluloid Corporation (formerly started by Hyatt) suffered from new competition. The Dreyfuss Brothers had gained controlling stock in Celluloid sometime before 1921. The company accumulated a one million-dollar debt in 1921 and lost over $1,500,000 in a three-year period from 1924-1926. In an effort to shore up their losses and insure their investments, stockholders put the firm under the wing of the Celanese Corporation of America, which made artificial silk by a chemical process quite similar to the formula for plastics, and which had patents on the manufacture of cellulose acetate that would help greatly in the production of Lumarith. This helped save the company and increase profits. In 1936, they declared a $2 dividend on the $7 preferred stock.

Decks of playing cards ($2.50 to $3.00 a deck) made from Lumarith helped produce this profit. Another novel application of Lumarith was for street lamps. Transparent sheets of Lumarith were much more resilient than glass; Lumarith did not shatter very easily. Lumarith pleated lamp shades sold extremely well.

In the 1930s, Celluloid suffered a steep and steady decline in popularity due to the invention of acrylics and urea plastics. By 1947 the Celluloid Division of Celanese Corp. ceased production of cellulose-nitrate plastics.

Japan continued to produce Celluloid items into the early 1950s. Once again, this was due to the fact that one of Japan's greatest natural resources was camphor. Today Japan still is noted as a manufacturer of celluloid items.

In 1941 Freeman, Taylor & Stanley of New York distributed Christmas tree ornaments, manufactured in Japan, made of nonflammable cellulose acetate. They were blown in a process similar to that used in producing glass ornaments and were priced on a competitive basis. Woolworth, Kresge's, and many other variety chains stocked them on their shelves. The New York display firm, J. M. Gordon & Co., produced fluorescent ornaments from Celluloid Corporation's Lumarith treated with special dyes to give a double pigmentation. These ornaments were priced from 5 cents for a simple icicle to $1.50 for an elaborate encrusted ball, six-inches in diameter. "Black lighting" units made by General Electric and Westinghouse were recommended for showing these ornaments at their best. A 100-watt black-lighting unit sold for about $30.00.

However, the Office of Precious Metals banned the use of cellophane and other cellulose materials for manufacturing such frivolous items as Christmas ornaments in November of 1941. OPM allowed companies a 60 day grace period in which to use up the cellulose on hand. Manufacturers quickly put some aside to use for Christmas 1942.

In a 1946 magazine, *Science News Letter,* DuPont chemists reported on the latest plastic, cellular cellulose acetate or "CCA" for short. They said it was "lighter than cork but stronger and a better insulator." It was uniform and capable of being tooled and shaped with woodworking tools. The new plastic could also be used in airplane instrument cases, furniture, lockers for frozen food, and even the core for light toys. DuPont chemists say the discovery of "CCA" was an accident. "When a cold slug of cellulose acetate clogged a molding machine, it was heated to help dislodge it. The result was the first cellular cellulose acetate." "CCA" was produced by heating a mixture of cellulose acetate with other materials under pressure. It came out as a mass of pinpoint size bubbles that became hard and rigid when chilled.

Thus celluloid's history slowly dissolved as hard plastic took its place in the manufacturing of dolls, toys, and other holiday items in the 1950s. But its use in previous decades provides for collectors countless impressive objects so very worthy of collecting and preserving for future generations to come.

Collectors need to be aware of several facts in the preservation of celluloid items. Sunlight will discolor celluloid, so it is suggested these items be kept out of direct sunlight. In addition, celluloid is extremely sensitive to heat and flame. Sometimes, celluloid items are a bit more grimy than desired. To remove this surface dirt, a very soft brush or cloth and some lukewarm water can be used. Use this procedure only with items in excellent condition. Be careful with a item that has greatly deteriorated. Water can do more harm than good. If water is used, be sure to remove it as quickly as possible with a dry cotton swab. A little pure vegetable oil can be employed to remove any residue from marking stickers placed onto celluloid items. Simply apply the vegetable oil and wait a few minutes. Then rub with a soft cloth. Never use nail polish remover and any harsh detergents to remove surface dirt or marking residue. Any serious damage should be referred to a professional restorer depending upon the value, scarcity, or desirability of a particular item.

Chapter 2.
Those Fabulous Holidays

The most fascinating celluloid pieces for many collectors are the holiday items. Halloween leads the way in rarity with Christmas, Easter, and other holidays falling in place behind. Since Halloween items are the scarcest, they command the very highest prices. Christmas items, with the ever-increasing number of collectors this festive holiday brings, also are quite high in price. In recent years, collectors have been drawn to Easter items as all Americans are starting to embrace decorating their homes for all the major holiday periods of the year. What follows is a chronological look at our holiday year and some celluloid items that could be used to grace any holiday decorating you may do.

Washington's Day

With February 22nd being the birth date of George Washington, the acknowledged "Father of Our Country," there were a number of different items produced in earlier decades to commemorate his birth. Very few celluloid items were made. Hatchets and celluloid buttons with his image are among those made. Any image of George Washington in a celluloid figure would be rare.

A Georgian Pair reminiscent of George and Martha Washington, jointed dolls, marked "Japan with cross in circle." $175-200 for pair.

St. Patrick's Day

While St. Patrick's Day was originally celebrated in Ireland, it now is more avidly celebrated in the United States with each passing year. Thus there has been more interest in such holiday items in recent years. According to historians, St. Patrick's Day was originally celebrated in Boston by members of the Charitable Irish Society on March 17, 1737. While primarily papier-mâché and cardboard items abound, there are some celluloid items available to collect. Most certainly, leprechauns and pigs fit this category. Pigs evoke the legend of "Paddy's pig" (a tale connecting St. Patrick or "St. Paddy" with his childhood days as a pig herder).

Rare jointed pig with clown riding and holding on to ears, marked "Japan." With coloration and design, it doubled as a St. Patrick's novelty item as well as a toy, depending upon the time of year in which this design was marketed. $165-190.

Leprechaun holding pig, "design number 1, Japan." $175-190.

4th of July

Four days after the actual signing of the Declaration of Independence, on July 8th, 1776, we Americans first celebrated this famous holiday. Now years later we continue this celebration. Interestingly enough, more and more Independence Day trees in patriotic colors of red, white, and blue are being seen every year across America. Therefore, July 4th collectibles have seen some steep price increases in recent years. Any image of Uncle Sam is counted as a very desirable item.

Uncle Sam "Made in USA" Trio of patriotic figures, earliest one on left. $225-275.

Halloween

While many Americans may consider Halloween a children's holiday, that has changed in recent years. October 31 is really the combination of "All Hallows" or "All Saints' Day" and "Samain." From the 1920s through the 1930s, Halloween parties were the rage. Countless magazine articles detailed decorations, party games, decorations, and food. Many of the celluloid items we have today can be traced back to their use at a Halloween party.

Most Halloween celluloid is extremely rare. That is obvious to any collector who attempts to locate some celluloid to put in a collection. Of particular scarcity are those roly polys. Most items were manufactured from the early 1930s up until the start of World War II. Salesmen would visit the stores to which they would sell and display blank samples. Then the store owners would choose the style they wanted as well as the paint colors. Thus we can account for the vast color differences seen.

Most of these were intended as party favors to be given to children at Halloween parties. Teachers purchased many of them—since they were such inexpensive presents—for their classes. One such teacher in southern Wisconsin remembered giving roly polys to better-behaved children and smaller standing figures to the others. Most of these were manufactured by the Viscoloid Company in Leominster, Massachusetts, as evidenced by the marking found on so many of these items.

Witch roly poly holding Ace of Spades card. $350-450. Extremely well-detailed and molded witch holding cat roly poly, marked "USA with circle." $575-650.

Black cat on pumpkin roly poly, marked "Japan ('circle cross' embossing). $500-575 (in perfect condition). Witch roly poly holding ace of spades. $425-525.

Early Punch (from Punch and Judy) roly poly, no markings, possibly German. $500-575.

Owl with open wings on top of smiling jack-o-lantern, roly poly, marked "Viscoloid." $700-775.

Owl figures, unmarked. $200-300.

Owl roly poly, marked "USA in circle." $365-425.

Two rare witch Halloween figures, left is witch driving car, no markings. $500-525; right is witch pulling dog in carriage. $600-700.

Extremely rare witch driving corn mobile car with broom at side, marked "USA with circle." $700-800.

Owl figures surrounding pumpkin girl with cat hidden behind figure, marked "USA with circle." The two on the right have pebbles inside. Left, $250-275; middle, $250-375; and right, $250-350.

Owl and Cat with inserted celluloid eyes, the inside of which have tiny black balls to make noise, marked "Japan." Owl marked with "cross" and cat marked with "flower in circle." Owl, $225-325; Cat, $150-250.

Later witch, marked "Made in USA." $175-180. Cat on pumpkin, celluloid ring to stabilize figure on bottom, marked "USA" in circle on back. $550-660.

Black cat, "USA in circle." $200-250. Girl carrying pumpkin candy container ready to go trick or treating, no marking. $300-350.

Skull meant to go in a flashlight while trick or treating, no marking. $175-200.

Witch candy container, of later vintage (close to transition period where celluloid turned to a more durable plastic material). Removable head to insert candy in body cavity. $225-260.

Thanksgiving turkey figures created through three separate molds, no markings. $100-125.

Jointed Pilgrim children dolls dressed in crepe paper clothing on cardboard disks. $65-80 each.

Thanksgiving

Unfortunately Thanksgiving has been lost in recent years between the holidays of Halloween and Christmas. First celebrated in 1621, it continued to be celebrated on somewhat an irregular basis until George Washington made this day an official holiday in 1789. In earlier decades, Americans decorated their homes profusely for this occasion. However, Halloween has become a decorating mania. It seems we now have little energy left to decorate any longer for Thanksgiving. But Americans still assemble each year for Thanksgiving dinner even though we do not decorate as we did in previous decades. The most traditional symbol of this holiday has been, and continues to be, the turkey.

Christmas

Christmas is the most popular and celebrated of holidays in the United States. As early as the 1840s, Americans started the tradition of decorating a Christmas tree. With the tree customs came the practice of decorating the entire home with Christmas items. From wreaths to Santas to snowmen, Americans filled their homes during this holiday.

Thousands upon thousands of celluloid items were manufactured for the Christmas holidays. Candy containers, ornaments, standing figures, dolls, cards, advertising objects, and so many other types of goods were produced for Americans who had an insatiable appetite for such items, especially those which were more inexpensive. Celluloid was readily available and quite low in price for those who were on a tight budget. Much was also purchased by neighbors, teachers, and distant relatives as gifts for the little children on their lists.

While standing Santa figures do abound, there are some real rarities among them. Christmas lights and roly polys are at the top of the rarity chart. Roly poly toys are extremely rare today as children playing with these fragile and easily dented gifts often destroyed them within a week of receiving them from Santa Claus. Many of the roly poly toys found today were store samples or given to adults who treasured them for their appearance and never dared to play with them. Ballast in the bottom—either a metal weight, sand, or pebbles—was used as a to keep the roly poly upright.

Two roly polys, marked Japan. Left, $200-225; right, $140-160.

Two roly poly figures, front view, index finger held to beard which reveals a message meant for a little boy or girl, "I have a surprise for you." Left, 150-165; right, $175-190.

Back view of Santas display the doll each is holding as a surprise.

Five roly polys, marked "Viscoloid." $150-175 each.

Two rare Christmas related roly polys, Santa in front of a house and a decorated tree, both "Viscoloid." $225-250 each.

Santa delivering presents was a perennial theme prevalent in many of the different figural toys and figures created for the American market. Very inexpensive in price, many of the items below were given by neighbors to children they didn't know so well or by teachers to their students at Christmas as tiny presents. Transportation items are always of interest not only to Christmas collectors but also to train, car, toy, and other such collectors.

Four cars with Santa delivering presents and trees, all marked "Viscoloid". $200-225 each.

Extremely rare train laden with toys and Santa acting as conductor, "Viscoloid." Trains were not often produced, companies favoring sleighs, sleds, or Santa on skis. $240-265.

"Viscoloid" zeppelins, in pink and green shades with Santa and toys below ready to deliver. Zeppelins are especially rare and difficult for collectors to locate. $350-400 each.

Two varying styles of Santa riding in sled, both "Viscoloid." Santa: $95-110. Santa with tree on sleigh front: $135-175.

Celluloid Santa on tin tricycle, marked "Suzuk" trademark "Patent No 424637." $150-165.

Santa on motorcycle with sack of toys, Japan. This is a rare item. $325-360.

While Santa in a sleigh with reindeer was the most common type of toy figure produced by the Japanese, American, and Germany producers, there are some rare examples which command some very high prices. Wind-ups and mechanicals are of particular interest.

Santa and sleigh with reindeer in original box, Japan, Registered Japan, USA, India, and England. $165-175.

Two miniature Santa and reindeer sets, one with one reindeer and other with two reindeer, all marked Japan. $85-95.

Reindeer pulling sleds laden with toys, all marked "Viscoloid." $125-135 (two on left); $140-160 (three on right).

Three Santa and reindeer sets, one with three reindeer, $165-180; one with two reindeer, $125-135 and another with one reindeer, $110-120. All made in Japan.

Celluloid Santa in sleigh. Gold glitter flocked. Early 1930s, Japan. $95-110.

Two deer pulling Santa on metal sled, all mounted on snow-flocked base, Japan. $75-80.

Windup celluloid Santa and reindeer on metal sled with bell in back. The bell rings when wound. Japan. $175-190.

Large celluloid sleigh with Santa and two reindeer, early 1920s, Japan. $100-130.

Reindeer pulling sleigh with Santa, all original, and entire piece is made from celluloid, Japan. $75-85.

Four-inch reindeer and sleigh, no Santa, all celluloid. $20-30.

Windups celluloid Santa and sleigh with original key. All including sleigh, reindeer, Santa, and whip are made of celluloid. $300-325.

Standing figures are quite common, but when made of celluloid they usually command some very high prices. Prices indicated are only for items of 100 percent quality. Dents, breaks, dirt, or a repaired figure would be much lower in price. While Santas with baskets or sacks of toys are the most common, those with tennis rackets or in different poses are extremely difficult to locate since very few were originally manufactured.

Four Japanese Santas all bearing lanterns, Japan. $75-100 each.

Assortment of six Santas in various poses, all "Viscoloid." Santa in the chimney, $125-135; Santa with "Merry Christmas," $125-135; the two Santas with holly mounds, $125-135; and the remaining two Santas, $75-100 each.

Assortment of Santa figures, all of which are carrying a basket of flowers and a pack over their shoulders, Japan and United States. Three with black boots, $100-110; one with brown boots, $130-140.

Four Santas waving with right hand and holding lantern in left hand, all marked Japan. $75-110.

Two large (six-inch) Santas, one with Jack-in-the-box and sack marked "Japan" $75-85 and one with open sack of toys and lantern marked "Made in Occupied Japan." $100-125.

Unusual Santa with tennis racket, marked "Japan." $165-175.

Two Santas carrying pack of toys over shoulders, both made in Japan. $140-165 each.

Two Japanese Santas, 1930s, one with a black doll and the other with a teddy bear peeking out of the sack. All definitely with the Japanese-style face. Left Santa, $110-120; right Santa, $95-100.

Two Japanese Santas, one with a heavy cap with a white band and the other is a more traditional Santa with pack on back, but with a more American appearance in the facial characteristics. Left, $75-85; right, $140-160.

Marked "Irwin—Made in USA," two simple later pieces of celluloid acetate in white with red painting and no other details. Later pieces were merely white material air brushed red with precious few details added. $75-80.

Two elaborately molded and painted Santas, marked "Made in USA." Early examples molded with great detail, air brushed, and then painted by hand for fine detailing. These two Santas date from the early 1920s. $225-280.

Three later American "Viscoloid" white figures, with simple air brushed red and green trim. $50-65 each.

Eight-inch Santa, Cross in circle mark, Japan, holding finger up to beard indicating "I have a surprise for you." $145-160.

Ten-inch Santa holding lantern, with pack over back, more European in appearance, marked Japan in back. Slight dent in head, which lowers value. $175-200 (in perfect condition).

Back view of Santa revealing a toy doll for a good little girl.

Three-inch "Made in USA" Santa, possibly Viscoloid, in tomato shade of red with dull painted features, carries basket of flowers, 1920s. $35-45.

Three-inch Japanese Santa, green pants which match green twigs he is carrying, yellow pack over shoulders, unusual detail. $85-95.

Two miniature Santas, two and one-inch heights, both marked "Japan." $45-60.

Toys of larger size and earlier vintage have attracted some cross-collecting interest from various sectors of the collecting public which has driven prices to great heights. Earlier toys in pristine condition command the highest prices.

Celluloid face on windup Santa skater who skates while waving his hands, Japan. His entire body moves forward and back while his right foot remains on the ground. 6" tall. $600-660.

Very rare Santa candy container with a squeaker, a celluloid head, wooden feet and a draw string in back for the candy container, German (Maxine Caldwell collection). $625-670.

Windup Santa, the head of which bobs up and down when wound, marked "Occupied Japan." $265-285.

Jack in the box Christmas Santa which pops up when the latch is released, paper covered cardboard with celluloid headed Santa dressed in original suit. $450-500.

"Occupied Japan" Santa on top of reindeer barn. When wound, reindeer heads pop up and down and merry-go-round ribbons of celluloid revolve. $275-300.

While milk glass figural lights, and other such electrical lighting, abounds in the collecting arena, "working" celluloid lights rank at the top since they were extremely flammable and often were damaged beyond belief upon one lighting of the tree on Christmas day. Thus any example found today is extremely rare and the value seems to rise even more with each passing year. The bulbs were of 3.5-3.7 voltage and had the C6 bases which were lit in series; thus if one bulb burned out, the entire string would go out as well.

German-marked Christmas lights, into all of which a light was inserted from the back or bottom of these celluloid lights. Extremely rare as heat from bulbs melted or destroyed fragile celluloid material. $225-250 each.

Japanese-marked cardboard Nippon Christmas light with celluloid Santa insert. $200-225.

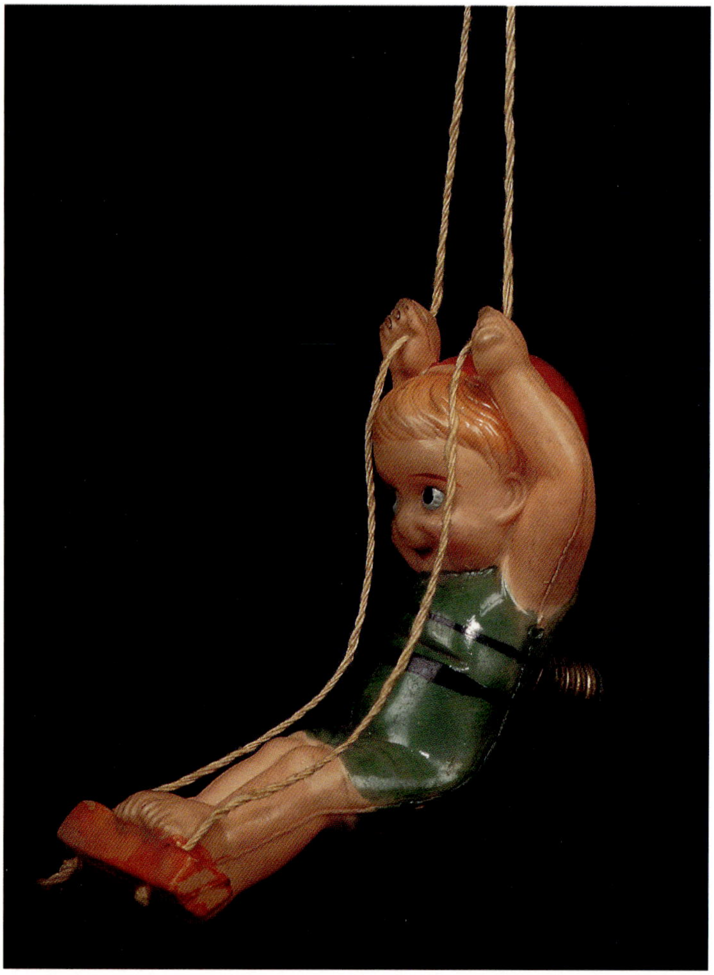

Japanese-marked Christmas light of girl in swing. $200-225.

Japanese-marked celluloid rooster light with light inserted in bottom. In pristine condition. Never used. $135-145.

Candy containers and tree decorations seem to go hand in hand, both being employed as tree ornaments. Many cotton figures, with netted candy sacks, and other such containers, doubled as tree ornaments and candy containers. Invariably, they were produced in Japan as the Germans and Americans never really did enter this sector of manufacturing.

Celluloid headed Japanese net candy container with felt hands and feet, 1930s. $200-225.

Celluloid headed Santa on top of net bag and fiber candy container, plaster hands and boots. Its rarity lies in the red fiber roping used at the bottom of the container. $175-200.

Three early examples of Japanese-manufactured net candy containers, celluloid bodies, composition hands and feet, net bodies. Left, $100-125; center and right, $135-150.

Cotton ornaments with celluloid faces glued onto cotton balls; one with twigs, possibly German, and other with single twig, Japanese. $125-150 each.

Later 1930s example of Japanese-manufactured net candy container. Drawstring in back. $135-145.

Later 1940s example of Japanese-manufactured net candy container. Drawstring in back. $135-145.

Candy celluloid basket with emerging Santa figure, marked "Made in Japan." $135-150.

Cloth-stuffed ball with boy's celluloid head, used as a Christmas tree ornament, Japanese, (Maxine Caldwell collection). $150-170.

While many celluloid animals and figures were extremely popular for using in Putz scenes and other Christmas scenes under the tree, on mantles, or tabletops; just reindeer and snow baby related materials are in this section. Animals and human figures were often sold in sets for Christmas giving. They were used in farm scenes while adult family members often recycled them (those that were not destroyed by playing) in later seasons in Putz and snow scenes. These will appear in the sections "Animals" and "Simple Human Figures."

Variety of reindeer, all manufactured in Japan. Those with red glass eyes, marked "Made in Occupied Japan," and any German-marked examples command the higher prices. The larger Japanese types also are slightly higher in price in relationship to their smaller counterparts. $15-45.

Early Admiral Perry roly polys, unmarked and attributed to Germany. $325-375.

"Tiny 1" standing snowbabies, marked "foreign with flower in circle," $50-65; and snowbaby on sled, marked "Hong Kong." $15-25.

Polar bears marked "Japan." $40-65 for solid bears. The jointed bear, however, is the rarest of those shown. $125-135.

Christmas novelties, advertising, and related material round out the types of celluloid items manufactured for Christmas giving and remembrances. Other Christmas advertising items can be found later under the advertising heading.

Trio of one-inch stick pins incorporating Santa. $25-50 each.

"Made in Occupied Japan" Santa rattle with eyelet at top for string, which could be hung from buggy. Red round Santa with white pack. $120-140.

Buttons Galore

From early on, celluloid buttons were manufactured in great quantities. The first celluloid lapel buttons made their appearance in 1892, not much different from current plastic buttons except that they had a closed back with a small button for inserting in the lapel buttonhole instead of the rolled edge holding a pin. The Colombian Exposition in 1893 introduced these buttons to a worldwide audience. The greatest impetus for production occurred in 1896 when Kinney Brothers (tobacco and cigarette manufacturers) flooded the country with a deluge of celluloid buttons that continued unabated for almost 15 years. The Kinney brothers inserted a "picture card" in each package of the famous Sweet Caporal cigarettes. Especially prolific was their output of "little actress photos." In 1896, they decided to delight their customers by placing a free celluloid button in every package of Sweet Caporals. At first, actresses were depicted on the small cardboard pictures. In a light brown or sepia, literally millions of them were distributed at the end of the 1890s. Other series followed: a series of national flags, a series of 48 U.S. state seals, and a long series of comic buttons. At first the comic buttons contained only text but soon were augmented with pictures of the Katzenjammer Kids, Mutt & Jeff, Happy Hooligan, and other famous comic characters of the day. A set of baseball star photos (heads) was issued in 1910. A tobacco historian has stated that Kinney brothers contracted with the button firm of Whitehead & Hoag for the delivery of one million buttons per day for a period of one hundred consecutive days—all of this for their initial order in 1896. Other manufactures followed suit including the American Pepsin Gum Co. and the New York firm of M. F. Miller.

When the celluloid button craze of two generations finally declined, many manufacturers went bankrupt or transformed their factories into other channels of production. Obviously, due to the hundreds of thousands of buttons produced, it would be impossible to list all the different subject material contained on these buttons. However, Santa Claus, Christmas theme-buttons, and political buttons are of extreme interest to collectors today.

Simple, common buttons with the bust of Santa sell from $50-75. Those with transportation, toys, and children illustrated on them command higher prices in the range of $125-150.

Celluloid covered Christmas blotter made by "Bastian Bros. Co., Rochester, NY. $40-55.

Easter

Easter really is a combination of the Christian commemoration of the resurrection of Christ with the ancient rites celebrating the arrival of Spring. The fertility symbols of the egg and rabbit are an integral part of this holiday; thus the proliferation of these two symbols. Ancient pagan tribes in Germany viewed the egg as the symbol of life's beginning. The yolk was associated with the sun and the white with the moon. The rabbit served as an escort to the Teutonic goddess Ostara, whose name was later altered to that of the holiday, Easter. Hiding and then hunting for decorated eggs is a custom which evolved in our country.

While the Easter collectibles market itself has not seen the huge price increases that Halloween and Christmas have seen, there is a keen interest in Easter celluloid items. It seems that collectors have recently discovered the whimsical, creative themes of these items. Therefore, Easter celluloid has risen sharply in the past two years. There is no doubt that this enthusiasm will not dampen, but only increase as additional collectors discover the charms of these toys, figures, and other Easter related celluloid items.

Pull toys: standing bunny, marked "Irwin," cellulose acetate, $125-135; miniature bunny on celluloid wheels, marked "USA in circle," $125-135.

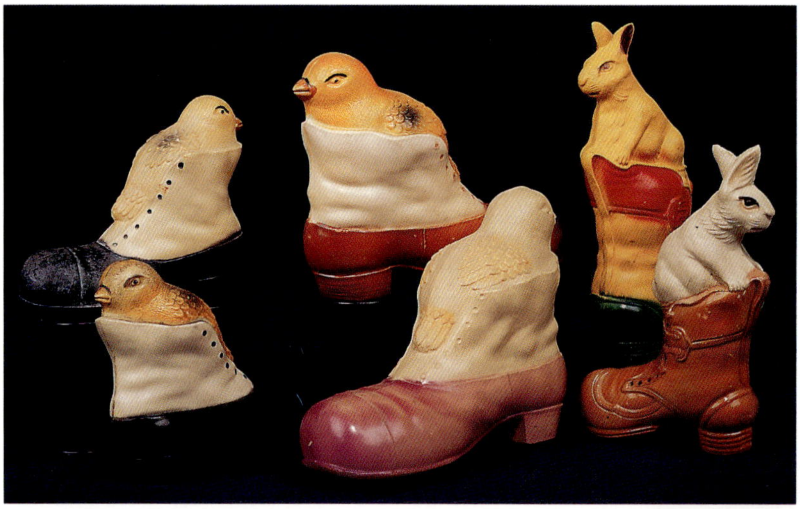

Chicks and rabbits in shoes, all manufactured in Japan and marked as such on bottom heels of shoes. $150-185 each.

Assortment of very detailed swan-related toys, most of which were meant to float on water. "Made in USA with circle marked on lower sides of swans." $165-195 each.

Animals in shoes, all marked Japan with cross in circle on lower heel of shoes. $150-175 each.

"Viscoloid" marked cars in similar styles, all with egg backs. $130-150 each.

"Viscoloid" marked cars, both were created from the same mold, but painted in different styles. $150-175 each.

"Viscoloid" marked cars, all made in America in similar styles. Very detailed molding with simple air brushed painting to finish details. $200-250 each.

Unmarked chick driving car of eggs, soft pastel colors, probably American-manufactured from molding detail of figure. $165-180.

Japan (cross in circle) cars, both from the same mold but one was converted into a pull toy with wheels and a front end celluloid string holder. Left, $165-185; right, $225-250.

Very unusual "Viscoloid" American-manufactured car filled with bunnies. Note the bunny riding on the running board. $400-425.

Unusual Easter-theme floating swans of Japanese-manufacture. Left, $95-105; right, $110-120.

"Viscoloid" marked swans with Easter chicks and rabbits. Note that the pieces on the left and right are from identical molds, but are finished in different colors. Very possibly, they were created in different seasons. American manufacturers varied the colors from year to year to attract different buyers for their Easter wares. Left, $75-90; right, $90-100.

"Viscoloid" marked boat with chick contrasted with pink rabbit in canoe, unmarked, but obliterated trademark is possible on canoe. Often the molding was rough. The trademark was often removed in efforts to smooth out the rough edges from molding imperfections. Irwin Cohn obliterated Viscoloid marks because they were in the molds when Cohn purchased the company in the late 1920s. Smoothing out the sides after taken from the molds is called "tickling" and each piece was hand done. The pink canoe is acetate. Top, $85-95; bottom, $75-85.

"Viscoloid" marked rooster pulling rabbit and egg-laden cart, both from the same molds but from different time periods. The example on the right is newer than the one on left. Left, $100-110; right, $85-95.

"Viscoloid" marked rabbit pushing chick in baby buggy, same mold but different painting. $110-120 each.

"Japan (cross in circle)" marked Easter rabbit pushing egg-filled cart and pink rabbit of similar mold marked "Japan (flower in circle)." Left, $95-105; right, $100-115.

Variety of animal drawn carts, all "Viscoloid" marked, all of which originally sold for 10 cents each. Most items of this type were priced from 5 to 19 cents each in the 1930s. $130-150 each.

"Viscoloid" marked carts, drawn by a variety of Easter-theme animals. Very small in size, they are more common and therefore are priced a bit less expensively. $85-95 each.

Assortment of carts, all manufactured in Japan. Many of these were almost direct copies of those created in the 1920s by American manufacturers. $100-135.

Japanese version of popular American-theme animals pulling carts of Easter toys and eggs. This theme was imitated in early 1930s by Japanese manufacturers who were quick to copy popular American-manufactured items. $95-105.

Variety of "Viscoloid" marked goose and rooster-drawn carts in soft pastel shades. Goose carts, $100-120 each; Rooster-drawn cart, $125-140.

Three types of animals are employed here to pull carts. Each example is marked "Viscoloid." Chick, $75-85; rabbit, $95-110; sheep, $150-165.

Extremely detailed wicker buggy, marked "Japan (cross in circle)." Rare, detailed molding in thin, fragile celluloid. $175-225.

Larger style "Viscoloid" marked rabbit pushing chick in buggy. $160-180.

Japan (circle in cross) roly poly with flower trimmed cap illustrating the detailed molding of the webbed "hands" of the duck and the feathers on the head. $300-325.

"Viscoloid" marked roly polys, all of which are extremely rare. As fragile as they were, roly polys were often broken. Coupled with this fact is the small quantity of these created for the American market. Since they were more expensive to create (due to the ballast in the bottom and the engineering design for motion), these toys were not produced in large quantities. $250-275 each.

"Viscoloid" roly polys, two from the same mold, but with different paintings. "Viscoloid" roly polys are decorated in contrasting soft shades. Each is subtle with light pastel shading, pink for girls and blue for boys. $150-175 each.

Variety of "Viscoloid" rabbit roly polys. Left to right: rabbit with eggs in basket on back, $155-165; brown rabbit on pastel egg, $140-160; miniature rabbit, $100-125; and rabbit with basket on arms, $145-

"Viscoloid" roly polys in vibrant detailed colors. Rabbit on ball (an end of the day piece), $285-300; bunny in pink suit with bow tie, $165-185.

Four Easter-related roly polys, all marked "USA in circle." Left to right: duck breaking out of egg, $75-85; chick with top hat, $150-165; standing duck in red vest, $175-200; and chick with tie and vest, $165-175.

"Made in USA" marked rabbit roly poly shapes in unusually whimsical themes with an inordinate amount of detail. Very prime examples of elegant Easter toys. Two on left are from identical molds, but painted in different shades. Left: $350-365 each; right, $175-200.

Three "Viscoloid" roly polys in egg shapes. Note the cracked egg with the emerging rabbit or chick. $200-225 each.

"Viscoloid" rabbits in hats with pebbles inside for the amusement of children, all from the same mold but with different paint details. Advertised both as rattles for children during the year and as toys at Easter, these rabbits enjoyed great popularity. Note the wonderful painted details. $300-325 each.

Chick breaking out of egg roly poly, "Royal Japan with fleur de lis" marking towards bottom of duck. Filled with sand and pebbles as ballast, this toy was extremely popular with children who loved the bobbing motion of the head when set into motion. $250-275.

Two "Viscoloid" roly polys in egg shapes, but more unusual in design. Elongated eggs were not often employed in molding as this particular design tended to have rough mold marks, which were difficult to disguise. Left, $375-385; right, $400-425.

The earlier piece with the red trimmed hat is marked Japan (cross in circle) and the other pastel shaded piece is unmarked (but obliterated "Viscoloid" marking is possible). The Japanese version is much lighter in weight than its American counterpart. Left, $260-280; right, $270-290.

Unusually shaped roly polys, marked "USA in circle." Chick riding rabbit on left, $370-400; and two rabbits facing each other, $300-325.

Rabbit candy container in celluloid cabbage head. Tiny celluloid head lifts off celluloid cabbage head to reveal a small waxed-paper cup originally filled with tiny pieces of candy. Chicks riding on top of a rooster in a wicker basket with great detailing. Both were manufactured in Japan. Rabbit container, $450-475; rooster, $275-300.

Unusually vibrant colors were used in this swan with the stocking cap, marked "Viscoloid" and the goose with the hat marked with a Japanese manufacturing trademark-design 4. Left, $140-155; right, $90-100.

Swan with muff, unmarked and swan with spectacles, marked design 6-USA. $145-165 each.

Assortment of standing rabbits, all made in Japan. Many of these same moldings were used for Christmas lights. If used as lights, the backs of the rabbits contained an insert for the electric light. They also doubled as toys and baby items at other seasons of the year. $100-175 each.

Japanese-manufactured candy baskets, the more unusual of which have rabbit and chick tops. Most of these appeared in children's Easter baskets and contained tiny amounts of rabbit and chick shaped candies. Three with heads on top: $180-200. Two with painted and embossed designs: $115-125.

Three later Easter figures, "USA with circle." $90-120 each.

Set of four Easter characters, marked "USA with circle." $90-100 each.

Freestanding bunnies, all marked Japan, the most unusual of which has separate floppy ears. $80-125.

Intricately detailed bunnies, all marked Japan. Left to right: rabbit with carrot in mouth and chick in hands, $180-195; rabbit with carrot and bow tie, $150-165; rabbit with basket tied to his back with red-painted ribbons, $140-155; and rabbit with yellow chick in hand, $200-215.

"Viscoloid" rabbits with golf clubs in two sizes along with "Viscoloid" rabbit with gun over shoulder. Red vested rabbit with golf clubs, $160-175; rabbit with navy vest and golf clubs over back, $185-200; and rabbit with gun over shoulders, $190-200.

Two monochromatic bunnies, marked "Made in USA." Left, $70-85; right, $85-95.

"Viscoloid" rabbits in varying sizes and designs. Three on bottom: $100-140. Two on top: 150-165.

Detachable-headed rabbit and chicks. All marked "Made in USA." Rabbit, $200-225; chicks $190-200.

"Japan (cross in circle)." Two figures: a detachable-headed rabbit on left originally intended to be a candy container, $240-255; Mrs. Rabbit with basket of vegetables, $200-220.

"Made in USA" figures, both of which have detachable heads in larger 6-inch sizes. Chick with open beak: $240-255. Mrs. Rabbit dressed in kitchen apron, $265-280.

Very brightly painted "Viscoloid" figures with very detailed molding. Chick with red scarf, $200-210. Rabbit with checkered cap, $210-225.

Japan-marked (cross in circle), very thin, fragile, and detailed figures with pebbles in the bottom to allow them to doubled as rattles. Chick, $135-145; rabbit, $165-180.

Very detailed "Japan-marked with cross in circle" figures featuring wonderful coloration and mold detail. Left to right: duck with scarf, $200-220; chick in dress and hat, $190-210; chick in dress, scarf and hat, $180-200; and chick dressed in formal wear with top hat, $180-210.

"Viscoloid" rabbit ... probably a transitional piece due to the yellow color and reddish air brushed red finish detail. This piece could be more into the plastic era than in the celluloid age and is very possibly cellulose acetate. $85-95.

Two "Made in Japan" figures: one is a jointed early piece constructed of extremely thin celluloid, $145-165; the other is an accordion playing standing rabbit, $150-185.

Three miniature rabbits from the transitional period with very little coloration and detail. American manufactured. $70-85 each.

Easter duck adorned with a sailor's hat and great detail. Marked "Viscoloid." (Maxine Caldwell) $150-165.

Jointed leg duck policeman, "Viscoloid" marked, five-inches in height and of a heavier celluloid. $225-250.

Easter celluloid head Japanese net candy container with composition hands and boots from the 1920s. Most of these pieces were created at Christmas and were given as souvenirs to children born in the Easter season. $165-195.

Chapter 3. Toys and Play

Things for Children

Some of the very first objects produced from celluloid were toys. Manufactured as early as the mid-1800s, baby toys and dolls soon were widely distributed as early as 1880. Prior to 1900, Germany dominated the toy market and it was Germany that produced the first celluloid dolls for Americans in 1878. Several years later, John Hyatt started producing dolls in America from scrap material under special patents granted to W. B. Carpenter in 1880 and 1881.

The very first celluloid boats, swans, and other such items were advertised as toys to float in bathtubs for children. They were blow molded from sheets of cellulose nitrate. Two 0.0015-inch thick sheets were placed together in a heated mold. Compressed air was then blown through a needle into the mold, forcing the heated material into the contours of the cavity. This combination of pressure and heat bonded the two sheets together at the points of contact, creating a single unit. Once the mold cooled, the toys were moved, trimmed, and hand decorated. The United States could not compete with Germany's experienced tool and die makers and its vast labor force (primarily women) who were willing to work for far lower wages than Americans. Thus prior to World War I, Germany was the principal manufacturer of such toys. However, during World War I, Japan with its ever-increasing manufacturing know-how, managed to take over Germany's dominance in this field. Thus, Japan became the main celluloid toy supplier for Americans in the period just after World War I. Their success was due to their vast resources of camphor and their low labor costs.

America soon began to explore this market as well. The Tennessee Eastman Corporation led the way in 1932, when it introduced Tenite, which in essence was cellulose acetate. Cellulose acetate was more flame resistant as well as capable of being injection molded. The first injection molded toys of cellulose acetate were manufactured in 1934 when Germans invented such a machine. The Dillon Beck Mfg. Company of Hillside, New Jersey, first successfully molded cellulose acetate into toys for the American public. According to Bill Hanlon in his book *Plastic Toys*, the earliest documented injection molded cellulose acetate airplane was illustrated in the October 1938 issue of *Modern Plastics Magazine.* In this magazine, there was an ad for Lumarrith manufactured by the Celluloid Corporation illustrating a "DC-3, which was molded by Kilgore Mfg. Co. of Westerville, Ohio."

Unfortunately celluloid was very fragile, flammable, and easily punctured. This accounts for the rarity today of such items. Celluloid items were also made for a very short period of time in Occupied Japan after World War II.

One of the drawbacks today to collecting such sets is the fact that many sets are incomplete or have suffered damage as a result of burns, warping, or staining from having come into contact with alcohol.

Dolls from Japan, Germany, and the United States

Cloth stuffed boy, five-inches in height with celluloid head, German (Maxine Caldwell collection). $200-220.

Attractive pair of six-inch celluloid dolls; girl with jointed hands and boy with glasses and matching scarf and cap, both made in Japan. $90-100 each.

Two celluloid dolls: a boy with a top hat and a pipe in his mouth and a girl with a wicker basket of apples, both are Japanese. Dating from the early 1930s, they were extremely inexpensive. Boy: $45-55; girl: $45-60.

Jointed twins in cloth blanket, unusual in that both twins have bottles in their right hands. Made in Japan in the early 1920s. $120-130.

Two jointed celluloid dolls: the doll in the red dress appears to have been re-clothed in a hand-knitted dress, $120-150; the standing doll is clothed in an early (original) blue dress, $180-200.

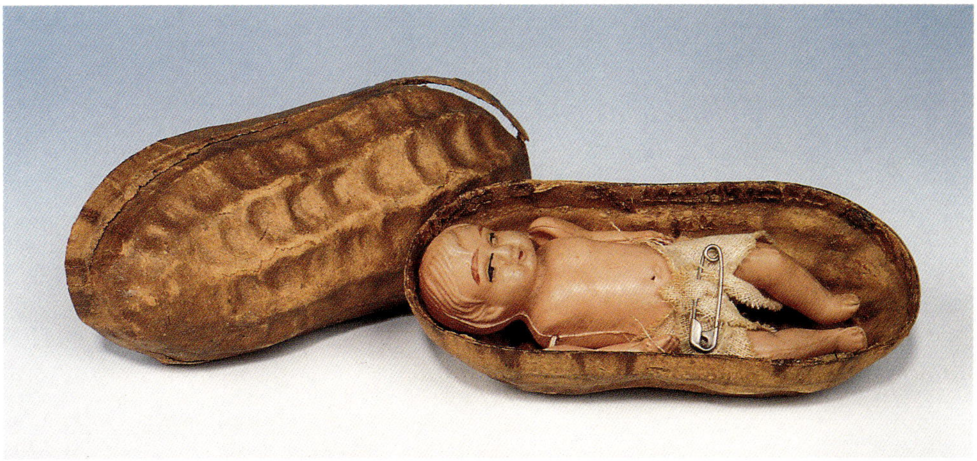

Jointed three-inch baby in diaper and sold in a cardboard peanut, made in Japan. $75-90.

Charlotte baby in bathtub (marked with "3" on back of doll). American-made. $45-55.

Jointed miniature doll, Japanese marked, 1 1/2" in height with its original gift box. Also used as a tree ornament. $125-135.

Kewpie groom doll in black crepe paper suit and white flower on lapel. Mum flower design on back of head, which is also marked "Japan." Kewpie dolls enjoy immense popularity with Kewpie collectors and therefore are more difficult to obtain inexpensively. $110-120; except for the Kewpie with the glasses, valued, $125-135.

Four more common Kewpie dolls, all marked "Japan." $25-65.

Unusual Kewpie dolls, one holding dog and other with pouch and shovel, both made in Japan. Left, $100-110; right, $75-90.

Black Kewpie doll, two-inches in height, made in Germany. Due to the ever-increasing number of black memorabilia collectors, these tiny dolls have escalated greatly in price in recent years and are considered rare finds. $125-145.

Elaborately cloth-stuffed Victorian dolls with celluloid heads, early but unmarked, used as Christmas tree ornaments (Maxine Caldwell collection). $225-250 each.

Jointed celluloid baby dressed in original clothing, German (turtle mark). $225-250.

Larger (8-inch) jointed doll, Japan, dressed in handcrafted, crepe paper sewn bonnet and long dress. $100-125.

Large (10-inch) jointed doll, made in Japan. $120-135.

Jointed (10-inch) doll with celluloid body, dressed in original clothing, German. $245-265.

Cloth stuffed doll with celluloid head, dressed in original clothing, early German. $245-265.

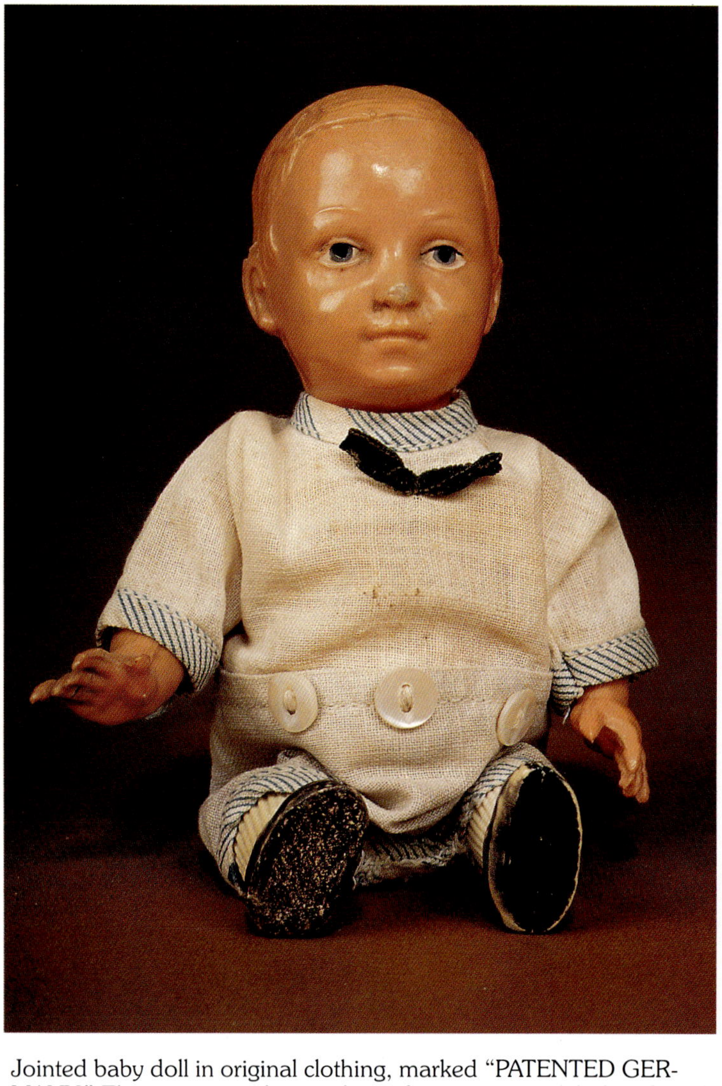

Jointed baby doll in original clothing, marked "PATENTED GERMANY." This piece was designed as a finger puppet. A hole is found in the back for a finger, allowing you to animate this jointed doll. $250-275.

Jointed figures, clown marked "Germany" on back and German jointed doll dressed in Victorian clothing and ribboned belt with metal buckle. Both date before World War I. Left, $265-280; right, $145-165.

Cloth stuffed dolls with celluloid heads, both dressed in original clothing, Early German. (Maxine Caldwell collection) $275-290 each.

Variety of jointed Japanese figures. The two dolls on either side of the soldier are beautifully molded and painted. The soldier was created as part of a military toy series. Left to right: early doll, $135-165; soldier, $45-60; red suited doll, $75-90.

Jointed doll in blue clothing and red cap, vibrantly painted blue eyes, marked Japan (design 25). $135-145.

Jointed doll in red suit tied with dark ribbon and green stocking cap, marked "JAPAN." $140-150.

Four boy football players, the largest of which was made in Occupied Japan while the others were simply made in Japan. The players without an eyelet at top each have jointed arms. $85-95 each.

Five policemen, all of which have jointed hands, all manufactured in Japan. The black suited policeman is the most difficult to locate. Red suited policemen, $75-90 each. Black suited policeman, $100-120.

Jointed figures from Occupied Japan, both marked on back as such. Some of the last items to be manufactured from celluloid. Doll, $125-135; bear, $100-110.

Jointed Japanese dolls, both marked "Occupied Japan." Japan was occupied by American military forces from 1945 to 1952. Some of the last celluloid manufactured dolls were created just after World War II. Left, $80-90; right, $100-120.

Dolls with attached celluloid baskets (could possibly be perfume holders): the boy on the left is marked "USA" with circle and the girls are marked "Japan" with cross in circle. $135-150 each.

A variety of rare food on tray items, all marked "USA in circle." Little girls use to play house with their "dollies" and this inexpensive celluloid food was so often used at playtime. $120-130 each.

A rare, double-sided early German doll: the head is a roly poly with different faces on both the front and the back of the doll. Left, $400-425; right, $375-400.

Back view of the German doll and the roly poly head, illustrating the different facial expressions.

"Spirit of St. Louis," New York to Paris imprinted on wings. This rare plane was a commemorative item. $350-400.

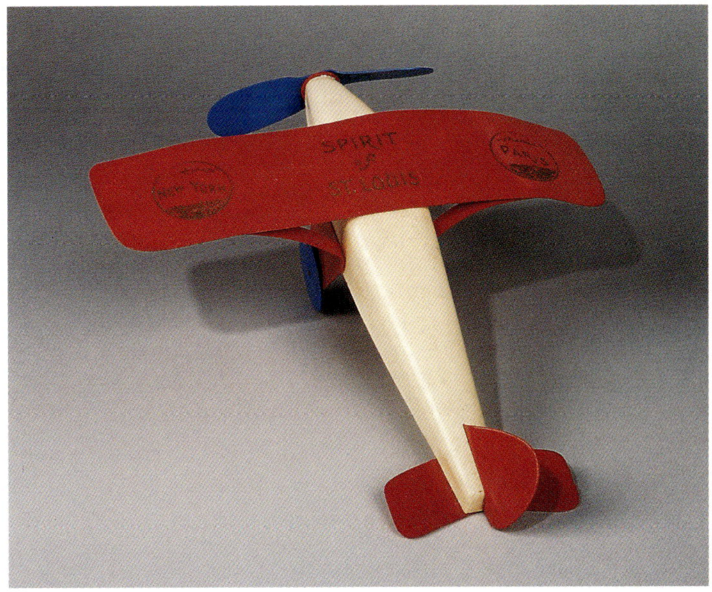

Top view of the "Spirit of St. Louis."

Roly polys with celluloid numbers and circus animals, marked "Made in USA." These are bridge table favors. They are extremely rare to find in any number as many bridge players took their single favors home, making complete sets very difficult to locate. $100-125 each.

Cat, dog, and clown: all of these might have adorned a Christmas tree. A tiny eyelet circle for string is found on each figure. All of these pieces were made in Japan. Since they were used as charms and ornaments, these figures are more often considered as toys than dolls. 3.25" to 5" high. Spotted dog, $75-85; clown, $105-115; Sandy, $90-100.

Dogs on seesaw balanced on a celluloid pin, which is subsequently inserted into a celluloid ball. $245-285; a clown riding horse, this piece was used as a blotter and is unmarked but the heavier quality and detailed molding make it a possible German piece, $165-180.

Assortment of baseball players, all made in Japan, those with rings were meant to be worn to the games and hung on the Christmas tree by fan. Also these 4" high figures are more often considered in the toy category than the doll category. $95-100 each.

Cup and Ball Tennis for Parlor or Lawn set in original box with Celluloid cups and four balls, manufactured by The Celluloid Novelty Co. of No. 313 Broadway, New York, dated 1890. Marked on the box, Trade Mark, "CELLULOID" Jan. 14, 1873... patented July 13, 1770. Reissued June 28, 1874. Caution— The word "CELLULOID" having been patented as a Trade Mark, any parties offering goods of any description under the name of "CELLULOID" will be prosecuted to the full extent of the law. $450-500.

Two baby windup toys, Japanese, crawl across the floor when wound by a key. Left, $165-185; right, $130-140.

Windup goose, Occupied Japan, counter balanced, weighted wings which flap and beak which opens and closes as wound-up goose waddles across floor. $240-260.

Mary (celluloid), when wound, pulls a lamb with a counter balanced, weighted head, metal wheels and a platform, Japan. $385-400.

Smaller version of the windup Mary in the previous picture, but the lamb is a complete figure with no bobbing head, Japan. $350-375.

"Made in Germany," six roosters in the original box advertising "celluloid toys" in Italian, French, and English. Original set in box: $425-475.

Close-up of rooster and paper label on box of toys.

Bunny and cart insert (animals and egg) are celluloid with a metal base and wheels complete with ringing bell in back, windup, Occupied Japan. Often included in Easter illustrations as it was originally intended to be sold as an Easter toy. $265-285.

"Obstinate Donkey" windup toy from Japan; clown moves back and forth as donkey stubbornly refuses to move. $365-385.

Windup Scotty dog which vibrates, moves tail and head when wound, Occupied Japan. After World War II, this was one of the last toys to be manufactured of celluloid. $75-85.

Earlier cat metal toy with celluloid head and tail which move as friction wheels propel the cat along the floor chasing a metal ball, Japan. $125-150.

Windup wooden cart with metal wheels which turn when wound, boy has composition feet and hands with a celluloid head, German. $600-645.

Jointed Japanese marked toy. Large clown of heavier celluloid construction with great detail. Marked "design 13." $125-140.

Celluloid man with target on metal ballast meant to go back and forth on a string, part of a toy, unmarked. $100-125.

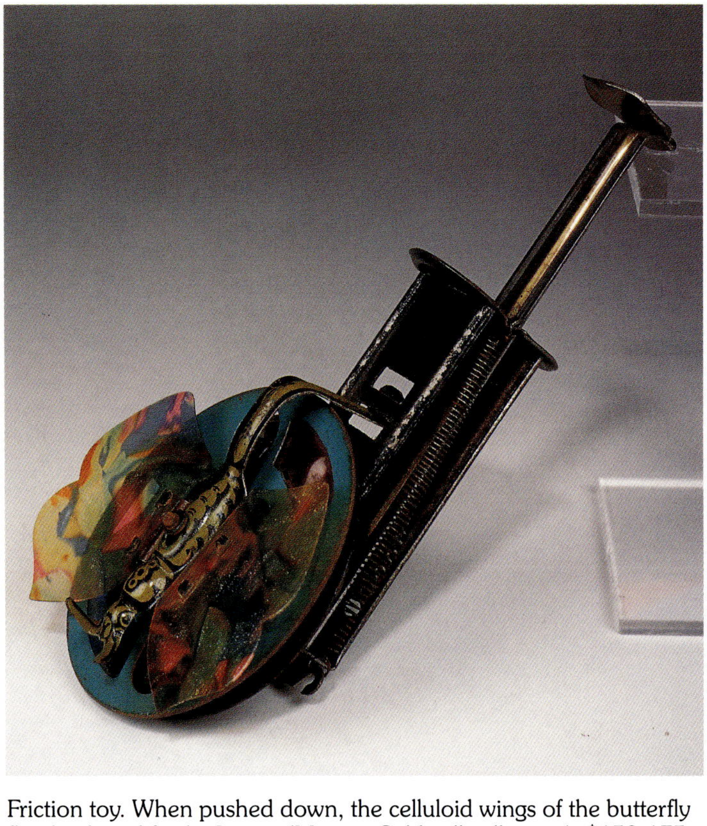

Friction toy. When pushed down, the celluloid wings of the butterfly flap back and forth, Japan. (Maxine Caldwell collection). $150-175.

Puppet on wooden jointed mechanism, celluloid head and original clothing, French. $285-300.

Windup black boy eating a watermelon slice; the child's arms move up and down as a dog bites his "behind," Japan. (Maxine Caldwell collection) $550-600.

"Barnum's Latest," celluloid container (1") with 1/2" camel, donkey, black boy, and two clowns. (Maxine Caldwell collection). $300-340.

"Dolly's Radio" toy in original box with cardboard radio, glass tube, and string metal headphones along with jointed celluloid doll, Japan. $245-265.

Bathtub fish, sold most often as toys for children to play with while taking their baths, all-Japanese manufactured. $10-35 each.

Dolphin, marked "USA circle." Also used in bathtubs more than any other place to entice children into taking baths. $85-95 each.

Ships, sold as bathtub toys, all American manufactured. Small ships, $65-70; large ships, $85-95.

Two ships, sold as bathtub toys, made in Japan. $50-60 each.

One ship, sold as a bathtub toy, marked Japan, design 21. $65-75.

Girl in red dress riding a horse, string replaced in hand and horse's mouth, marked "Made in Japan." $100-120.

Cat's head decorated with a ribbon collar on a musical cylinder with blue and gold dots, unidentified. Push down on the head and it plays. $150-165.

Three swans, all sold as bathtub toys, Japanese manufactured. Toys such as this helped to get children to take baths without fighting with them. The bottom two swans appear to be cellulose acetate. $45-50 each.

Cowboy riding horse, cowboy bounces up and down as horse vibrates, windup, made in Occupied Japan. $130-145.

Baby and celluloid bottle filled with white powder to simulate milk, American made (design 23). $145-165.

Father driving, mom in second window, and two children in back of this rare "Viscoloid" marked car. $250-275.

Large 6" buggy with jointed celluloid baby complete with brown human hair glued onto its head. Not marked, but the molding on the doll's hair point to a German-manufactured toy. $245-265.

Soldier riding horse with flag, entire item is celluloid, made in Japan. Part of a series of national soldiers with flags. $135-145 each.

Trio of celluloid buggies filled with celluloid babies, the rarest of which are still in original clothing, marked "Made in Japan." $175-190 each.

Trio of colored buggies, metal push handles finished with celluloid handles. Even the wheels are created from celluloid. Marked "Made in Japan." $175-190 each.

Pull Toys

Most inexpensive when compared to tin and wooden toys, pull toys appealed to parents and gift givers looking for a present, but unwilling to spend a greater amount of money. The bright colors and detailed molding appealed greatly to the children who received them as presents.

Pink celluloid lion on metal wheels, Japan-marked (design 22). $175-195.

Variety of wheelies, all made in Japan. All are beautiful, detailed examples of early 1920s celluloid toys. Small to large: $145-345 each.

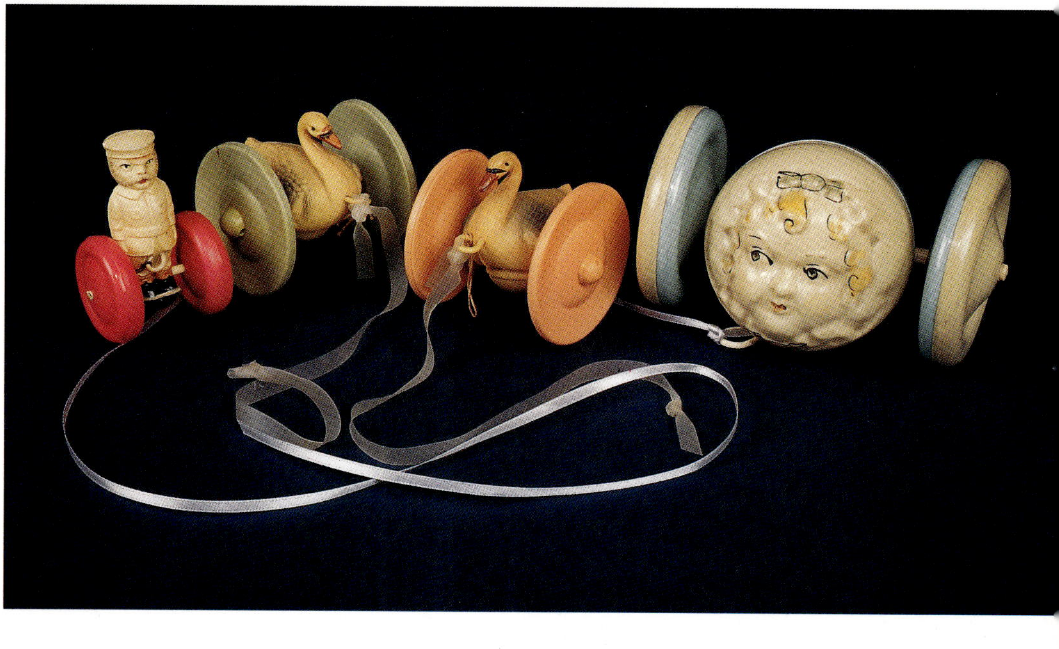

Pull toys with weighted roly polys, all marked Japan. Note the detail of the weighted roly poly center, which indicates an earlier manufacturing date. Most likely early 1920s. $190-210 each.

"Made in USA" platform pull toys with lions. $120-130 each.

Dog pull toy, 10" in length with a celluloid circle at end of the string. Pink wheels and coloration of dog itself point to an early 1930s manufacturing date. $145-165.

"Made in USA" platform pull toys with camels. Very small and simple in design, but very affordable for gift giving in the early 1920s. Note the celluloid platform on which the animals are glued. $125-150 each.

"Made in USA" platform pull toys with circus animals. More desirable because circus-theme items always appeal to various groups of collectors. $145-165.

"Made in USA" platform pull toys with farm animals. $135-145 each.

"Made in USA" platform pull toys with cows. $100-135 each.

"Made in USA" platform pull toys with pigs. $120-130 each.

"Made in USA" platform pull toys with varying sizes of buffaloes. $145-160 each.

Roly Poly Toys for "Good little girls and boys."

While roly poly toys have always enjoyed great popularity among children, those made of Papier-mâché and plaster were quite expensive. Since celluloid toys were so inexpensive in price, these toys appealed once again to those gift givers who wished to please their children, but just could not afford the very high prices of the other types. Most of these celluloid roly poly toys were sold in "five and dime" stores and distributed by Butler Brothers wholesale company who promoted their inexpensiveness, color, and detail.

Three rare roly polys, President Teddy Roosevelt, Buster Brown, and Dewey and Perry sitting on top of the earth. All are rare since they represent some well known figures of the period. Roosevelt, $350-400; Buster Brown, $325-350; and Admiral Perry, $425-450.

Unusual, well-detailed roly polys, marked "Made in USA with circle." Very detailed, early examples in very whimsical poses. Girl, $135-150; clown on ball, $165-180; and upside down clown on ball, $190-215.

"France" embossed roly poly, also of the penny toy category. Made in late 1890s. $200-220.

"German" embossed roly poly, of the penny toy category. Made prior to World War I. $165-180.

Roly polys, all marked with "H" on back. Possibly created by Dr. Paul Hunaeus, a revered German artist whose early designs were marked with an "H" while his later items were marked "PH." Produced in Germany. $130-150 each.

Unmarked roly poly gentlemen, all small and could fit in the penny toy category. $100-125 each.

German roly poly, marked "German" and appears to be of early vintage. $150-165.

"Nippon" marked roly poly, marked "design 8" on back. $150-165.

Assortment of penny toys, all unmarked roly polys. Made in USA, Germany and Japan. $125-160 each.

Miniature 1" roly polys, unmarked. But appear to have been produced in Japan in the early 1930s. Many of these were consolation prizes given at county fairs and carnivals. $35-50 each.

Soldier and clown roly polys, marked "Made in USA" and appear to be of World War I vintage. Soldier, $135-150; clown, $125-140.

Assortment of roly polys marked in blue ink "Germany." Are of later vintage. Unusual in set of five as illustrated. $125-140 each.

Punch (from Punch and Judy) roly polys, all unmarked. These are hard to find. $225-250 each.

Three bulldogs, roly polys, marked "Made in Japan." All are early 1920s pieces. $125-160 each.

Assortment of animal roly polys, all unmarked. All illustrate the whimsical quality and variety of animal toys produced in Japan and the United States in the 1920s just after World War I. Small to large: $125-200 each.

Cat roly polys marked with an unknown shield label, no country of origin is certain; however, they appear to be of early 1920s Japanese-manufacturing. $175-190 each.

Dog roly poly, marked "Nippon." $165-180.

Cat roly poly (blue base), marked "GERMANY." $200-225.

Variety of chickens and roosters, all marked "Made in USA with circle." These items were extremely popular in stores at Easter. $100-120 each.

Ram marked with blue stamped "Japan;" squirrel roly poly, embossed with "Japan;" swan stamped with blue label "Japan" on bottom. Ram, $65-75; squirrel, $80-90; swan, $45-60.

Rabbit, fantasy rabbit faced boy with ears, and alligator, early unmarked Japan. Rabbit, $190-210; fantasy rabbit boy, $240-260; alligator, $165-185.

Old West-theme toys

Perennial favorites of children were toys, which represented Indians, cowboys, and any aspect of our pioneer history. Especially popular were Indians in canoes and cowboys with guns.

Indian roly polys, marked "German" and "Japan." Two shorter German toys, $185-200 each; three taller Japanese versions, $125-140.

Two Indian in canoe items, the smaller model has a jointed head and arms while the larger example has jointed arms on the Indian chief, both marked "Japan." Top, $95-105; bottom, $50-65.

Two cowboys, both manufactured in Japan: the smaller cowboy is marked "design 9" and larger one is marked "Japan with Cross." Left, $125-135; right, $75-90.

Indian celluloid needle or match holders that separate, each with a tiny insert for some miniature object. These pieces have a celluloid Indian top balanced on the leaded feet base. Both are marked "Germany." Left, 200-215; right, $190-205.

Large eight-inch Indian wearing a feathered headdress, with jointed legs and hands, one of which holds a metal hatchet, patented "Japan." $225-240.

Buffalo, all marked "Made in USA." $45-50 each with larger animals on the higher end.

Simple, Plain People

All of the following are classified as "simple, plain people" only because they are of simple design and most do not have any jointed arms or legs. They could be included in the toy category as some were clearly intended for gifts to children and adults. Some were premiums while others were merely giveaways from businesses and companies to customers in appreciation for their business.

We know that many of them appeared at Christmas in "Putz" or "Village Scenes" under the tree or on tables. Numerous photographs of the time as well as magazine articles suggesting their use as such are evidence of this. Policemen would guard the roads filled with cars. Men and ladies would appear in front of Japanese paper houses and by bottle brush trees. These figures were so inexpensive that many Americans found them more attractive than the composition or lead figures produced during this period.

Larger boy with baseball, Jackie Coogan again, marked "design 10." Note both the wonderful positioning of this figure, he is getting ready to pitch the ball, and the clever tilt of his baseball cap. $195-225.

Two boys, images of the child star Jackie Coogan, marked "USA" in a circle." Both are exceptional in facial expressions and pose. $150-165 each.

A variety of poses are presented in all of the figures. These figures are unmarked. Left to right: policeman with gun in holster, $80-95; girl in cloak with fiddle, $75-85; little sailor boy, $95-105; and child with basket of flowers, $65-75.

A variety of Japanese-manufactured girls. $100-125 each.

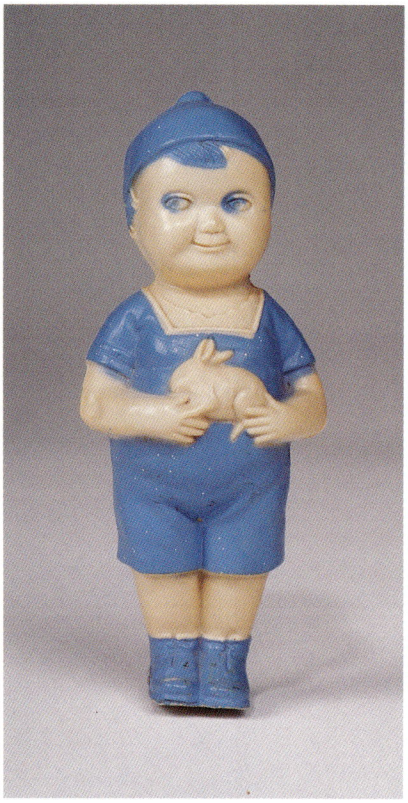

Little boy in blue clothing holding a rabbit of later cellulose acetate, marked "Irwin non-inflam." $75-85.

Variety of Japanese-manufactured clowns and boys. Left to right: boy with red top hat, $70-85; clown, $100-110; and boy with a sail boat, $55-70.

Two boys with bouquets of flowers, light version on left is German and later, darker interpretation on right is American. Left, $150-175; right, $125-135.

Girl with horn and girl playing violin, both from Japan. $100-120 each.

Two Japanese-manufactured girls, the girl on the right carries a with purse and the girl on the left, a suitcase. Left, $125-140; right, $110-125.

Girl nodders with strings, marked "Japan" with cross in circle. Very soft pastel shading with bright red trim. $120-130 each.

Large 7-inch Cracker Jack boy ("Sailor Jack"). Was he intended as a promotional item to advertise candy, a mail-away prize from Cracker Jack, or merely as a toy given to good little boys and girls? $225-250.

Black Memorabilia

While a compilation of simply dressed people continues, there is a group among them that has seen some momentous price increases in recent years. The numbers of black memorabilia collectors are rapidly expanding, escalating the prices for any and all items displaying African-American images.

These black minstrel characters have "design 14" inscribed on the back of each. These minstrels were often given as prizes at county fairs. Left, $265-290; right, 245-260.

Alligators with inserted celluloid headed pencils. Many of these were souvenir items with the city in which they were sold printed in black or blue ink on the top of the alligator's body. $125-165 each.

Pin cushion with attached celluloid black baby, marked "Japan." $175-200.

Black boy throwing dice, no marking. $200-215. Black jointed baby marked "ES Germany." $100-125. Tap dancer with cane, "W in circle and Germany" as marking. $190-215.

These black babies vary in sizes from two inches to seven inches high. All of these babies are made in Japan. Smaller to larger: $70-165.

Jointed arms give this black boy an added dimension. Marked with (design 18). $225-250.

Black girls, one with flowers and the other with a bowl of food with an inserted spoon, marked "design 17." $135-150 each.

Animals

Thousands upon thousands of these figures were produced over the years. Some were intended as part of a farm set, some were meant to be bathtub toys, and others were just inexpensive presents for children. Since they are so prolific, their prices are much lower than for other celluloid items.

Assortment of animals with metal feet, all Japanese manufactured. Many items of this type were originally sold as part of farm sets for little boys. Later, when the children were weary of playing with the toys or had grown up, their parents used them for Christmas scenes under the tree and on table tops. The three personality chicks on the left are valued at $125-135 each; the rooster and hen value around $120-140 each.

Chickens and rooster, marked "USA in circle." $65-80 each.

Mother duck and two ducklings, metal feet. $120-130 for each piece.

Pair of geese with inserted metal feet, marked "Viscoloid." $85-100 each.

Gray-striped cat with jointed head, marked "Juno" in circle and "GESG" and GERMANY underneath circle. $145-160.

Pink colored camel with beaded neck held in place with elastic string, most likely a children's toy. Made in Japan. $135-150.

Standing dove figure, in ivory tones, marked "Germany" at base of bird. Pre-World War I piece. $125-135.

Cat in shoe, ribbon around collar of cat, marked "JAPAN." Late 1930s squeak toy. $150-185.

Pastel colored ram, molded base of grass, made in Occupied Japan. $100-125.

Animals including a dog, rhino, bear, and others. All marked "Made in USA." Bears and dog, $20-30 each; buffalo, $45-50; rhino, $50-55.

Variety of dog figures, all American manufactured. $25-35 each except for the black and white spotted dog, valued at $55-60.

A variety of dogs, including some rare and unusual breeds. All marked "Made in USA." $35-45 each.

Various turtles and frog, all Japanese manufactured. $45-55.

Varying styles and sizes of elephants, all American manufactured. $45-60 each, depending on size.

Various animals, all made in Japan and marked with "Japan" (cross in circle). $25-35 each.

Irwin dog, marked "non inflam."and of later cellulose acetate composition. Produced in the late 1930s in the United States. $15-25.

Ram and two sheep, all marked "Japan." $25-35 each.

Swans with celluloid wings, marked "Made in USA." Celluloid wings make these figures much more rare than the single molded swan pieces. $35-45 each.

Cloth bodied black and white bear with a celluloid head, tied with a faded red ribbon, no markings. $100-120.

Jointed Japanese teddy bears, all marked Japan. All are post World War II. Smaller to larger: $85-115 each, depending on size.

Jointed elephants, marked "Germany." Unusual poses and excellent detailing. $180-195.

Jointed animals, marked "USA with circle." Rabbit, $115-130; duck, $100-115; dressed rabbit, $165-180.

Jointed tiger in large size, Japan "design 24." $50-65.

Jointed animals, marked "Japan with cross in circle." Duck, $130-140; chick, $145-160.

Jointed clown and dog, both marked "Germany"

Trio of cats, one with mouse, one with ball, and the other standing. All made in Japan. $60-75 each.

Three donkey nodders, all made in Japan. $75-90 each.

Desmond figure. "USA circle marked" Des. patent 72007, copyrighted by Dean's Ragbook Co. LTD. $145-160.

Nodders, swan and sheep, marked "Germany." Swan, $35-44; sheep, $25-35.

Five-inch red elephant with red trunk and filled with a hefty quantity of pebbles for use as a rattle. No markings but this elephant is representative of later American-produced items. $85-95.

Rare early German dog with muzzled mouth, embossed "Germany." $145-160.

Nodders, elephant and dog (with celluloid pole for head nodder), marked Japan design 2. Elephant, $95-105; dog, $125-135.

Felix the Cat, marked Japan (design 1), and another cat figure, unmarked. $200-250 each.

Cat on ball, "Fleur de Lis" mark on the cat's back, Japan design 10. A later Japanese piece dating from the post-World War II era. $125-135.

"Viscoloid" animals. Seal, circus elephant, pelican, and a rare end of the day marbleized camel. $35-65 each; except for the camel, valued at $120-140.

Two Japanese manufactured roosters: the one with dark feathers has a squeaker hidden amidst the feathers. $300-325 each.

Peacock on pillar, Marked "Japan (fleur de lis)" on back. Japanese manufactured. $100-125.

Wooly Lamb with colored curls and a red ribbon around the neck, "SK" marked on back. Japanese manufacture, of later vintage. $140-165.

This two-inch jointed monkey sports a top cap and cane. Marked "Japan with S in circle." $95-100.

Chapter 4.
Baby Items

This category is much larger than many collectors might imagine. With the birth of our babies, we have all bought present upon present to bestow upon each newly born child. Teething rings, rattles, tiny crib toys, and even grooming sets were popular presents from parents, grandparents, aunts, and uncles. Due to the fragility of celluloid, many of these items are scarce on the market today. However, a little diligent searching will provide some outstanding examples of artistic and creative objects.

Five teething rings, all Christmas-theme, American and Japanese manufactured. $150-175 each.

Teething rings, nurse, girl with baby doll, and rabbit carrying suitcase, all American made. $145-165 each.

An assortment of eight teething rings including orange, Easter egg, and double-sided baby face, most of which are American-made. $100-125 each.

Six teething rings, all circular in designs with dogs, cats, and storks, all marked "Viscoloid." $95-105 each.

Two teething rings with wooden beads strung with elastic string, both marked "American." $135-150.

Four teething rings with detailed molding including a clown, $140-165; an Indian with a guitar, $175-190; a child jumping rope, $135-150; and boy and girl cats holding hands, $250-275. All Japanese-made.

Three baby in wicker basket teething rings, all "Viscoloid." $200-225 each.

A variety of birds in rings and a rare monkey in ring, all American-produced teething rings. Bird, $95-120 each; monkey, $135-150.

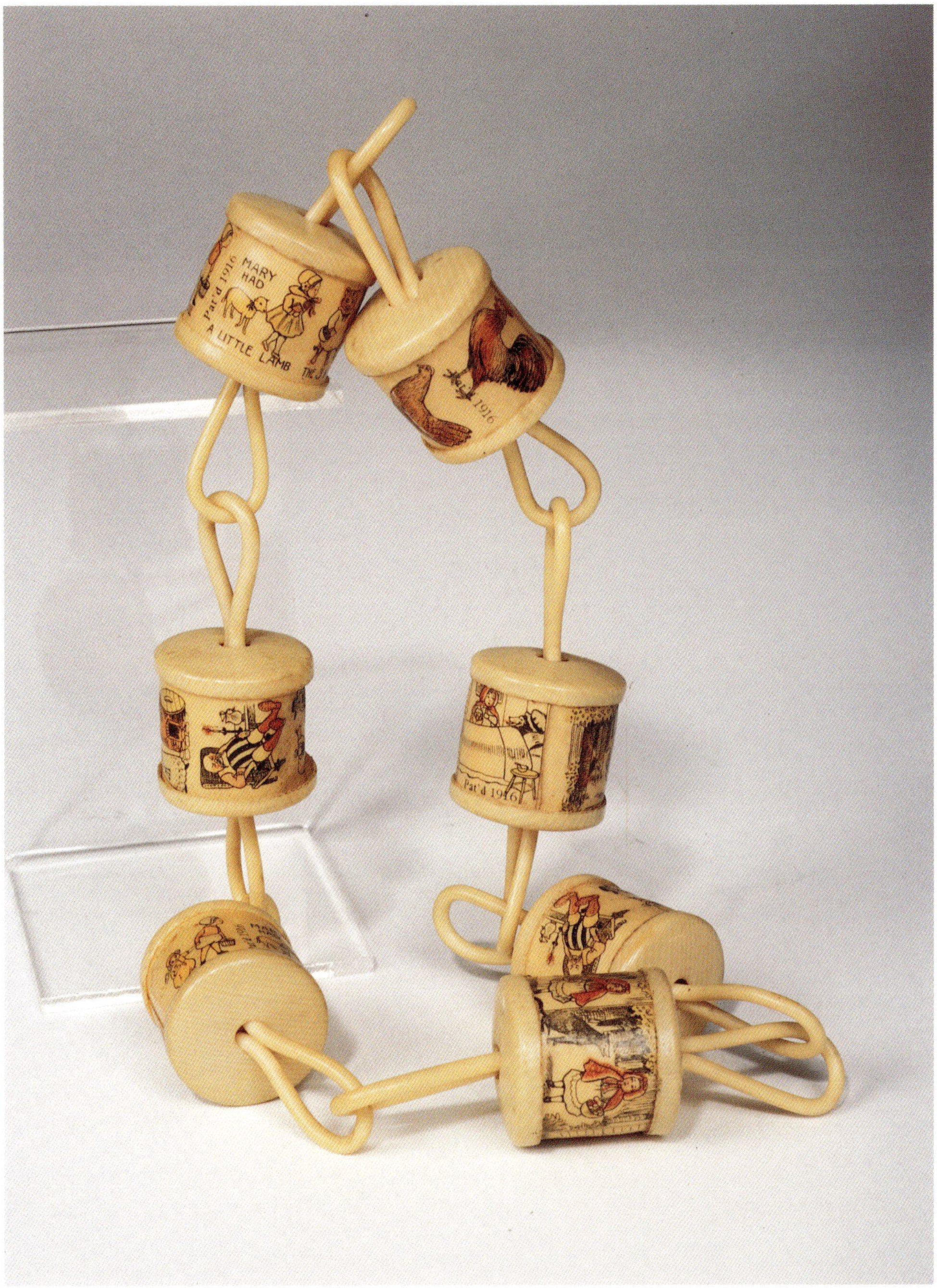

American-made nursery rhyme teething rings, including Jack and the Beanstalk, Little Red Riding Hood, The Three Little Kittens, and others, patented 1916. $425-470.

Close-up of one of the cylinders in the nursery rhyme teething ring.

Three teething rings including a red and white bell, a baby painted ball, and a double-sided baby cameo, all American produced. Bell, $65-95; ball, $125-135; cameo, $135-150.

Pink and green teething ring, American made. $85-100.

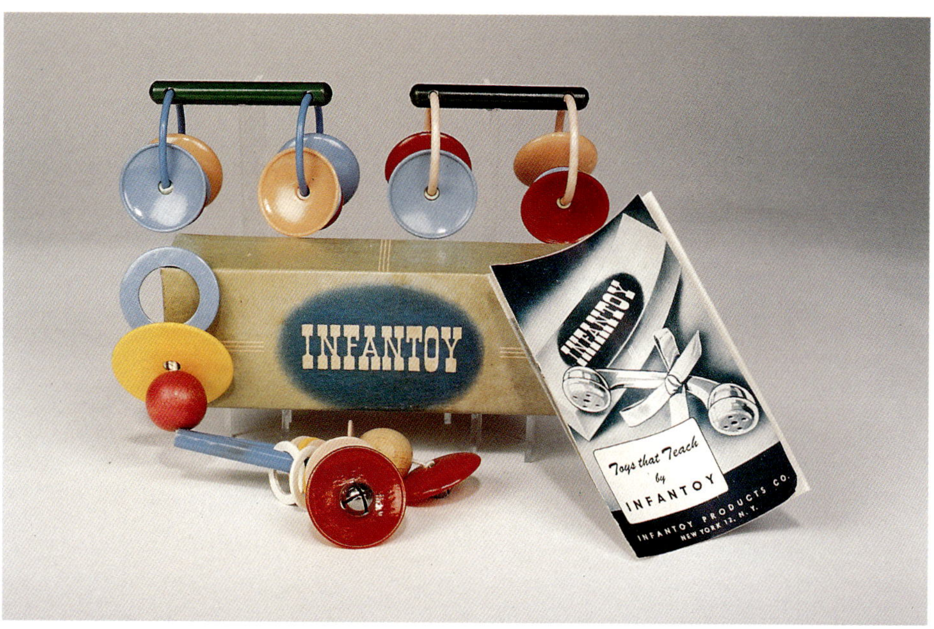

Various types of teething rings manufactured by INFANTOY Products Co., of New York, NY, in original boxes. $65-85 each.

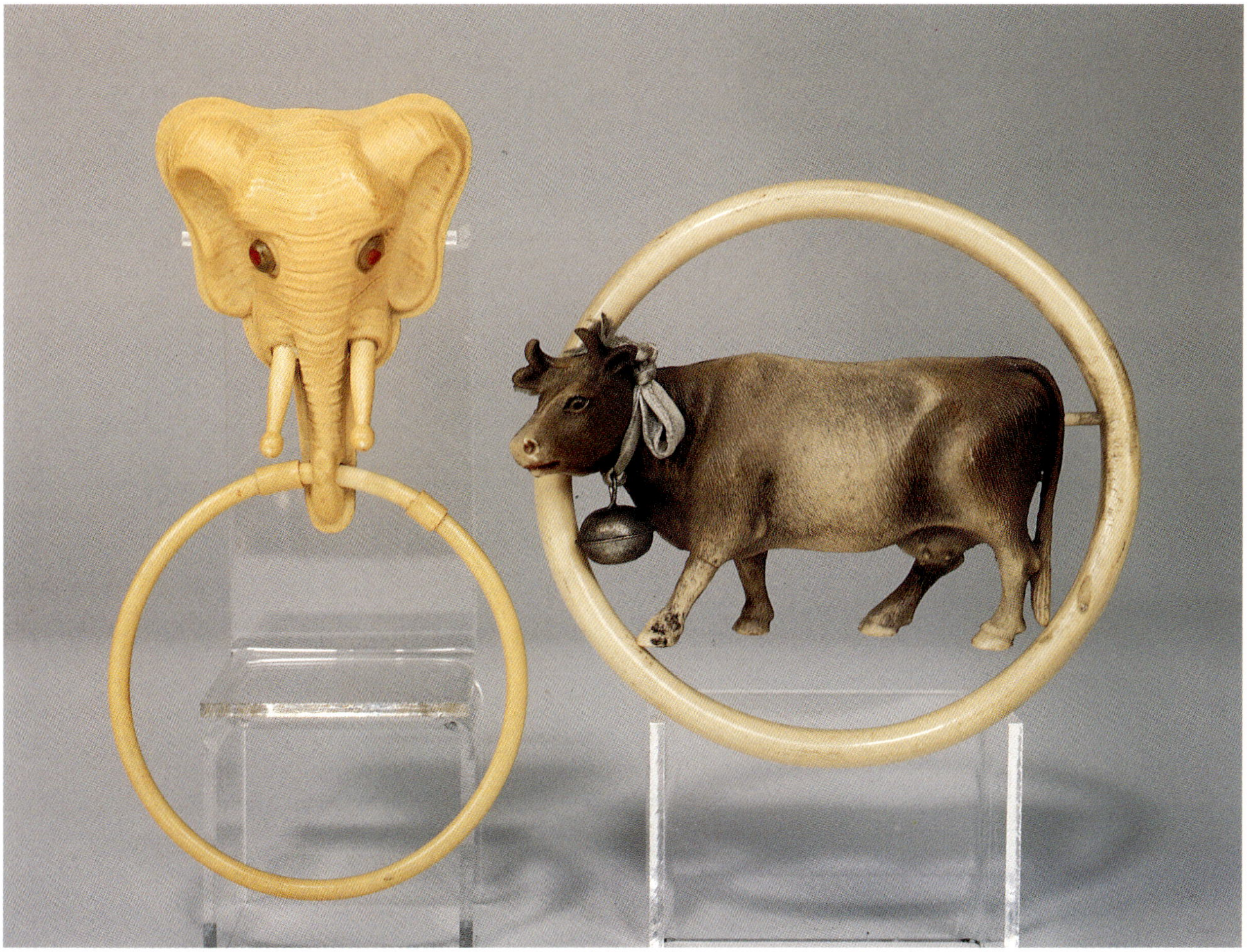

Elephant and cow teething rings, both American-made. $185-235 each.

Other baby items included rattles. Once again, their inexpensiveness helped to fuel an immediate interest in buying celluloid rattles. Their color, detail, and creative designs also helped to convince Americans that they *needed* more than one rattle for their child.

Four detailed examples of baby rattles including a boy in a hot air balloon, $175-225; a baby in a bathtub, $145-160; a standing child in a suit with a ball, $125-140; and a baby in an open flower, $180-195. All unmarked.

Four "Made in USA" marked rattles including a sailor with an anchor, $185-200; a cat in white opaque celluloid, $125-140; a cat in a transparent, amber flower, $130-145; and a baby in a flower, $110-125.

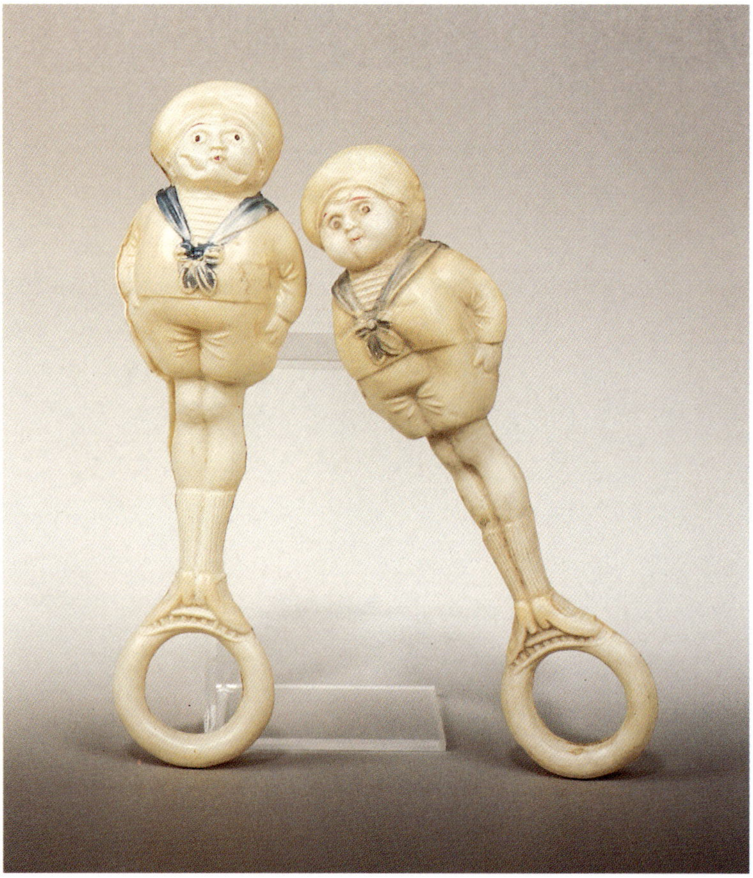

Two early rattles which date to the very early 1900s. Boys in sailor suits, marked with "X in square" on back. $225-250 each.

Four "American-made" rattles including a cat's head, a stork with a baby, a cat in a wicker basket, and a dog with a bow-tie. $95-125 each except for stork with a baby, $165-185.

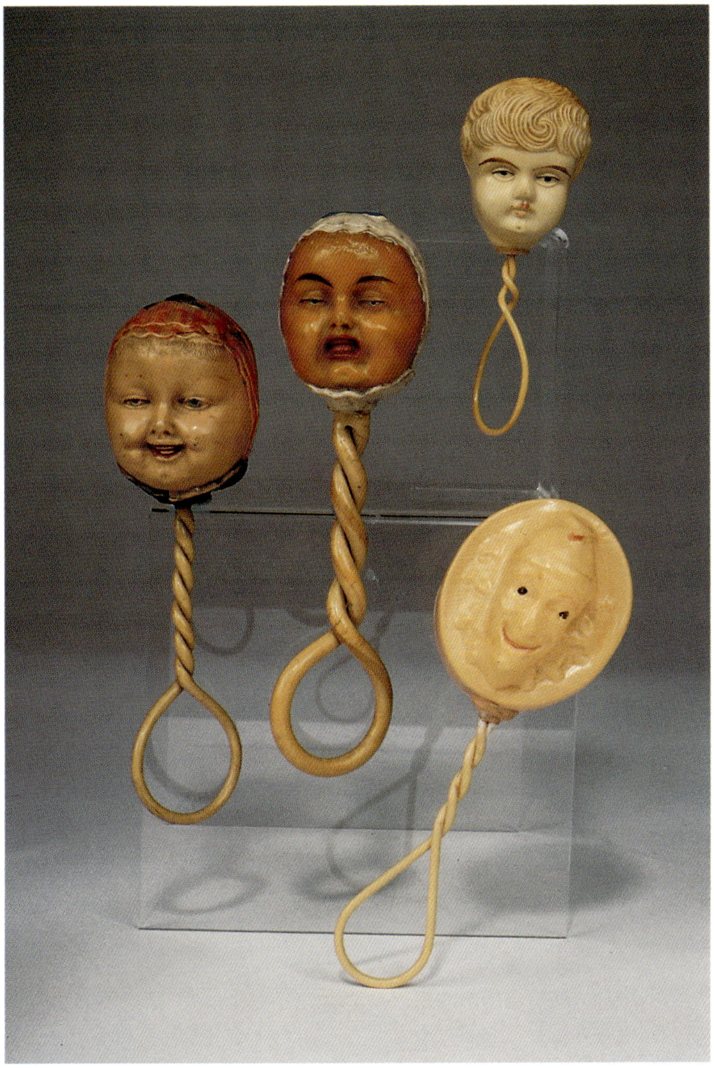

Four double-sided baby face rattles, unmarked (one side). $175-225 each.

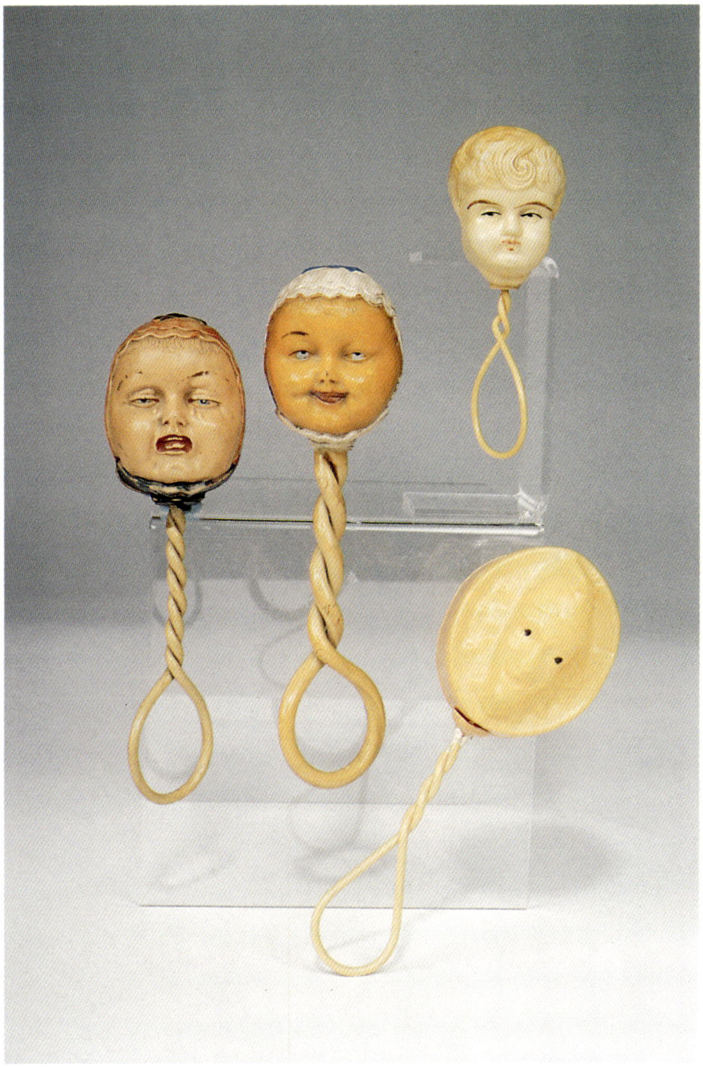

Four double-sided baby face rattles, unmarked (other side).

Seven double-sided rattles, unmarked (one side). More simple in design, yet whimsical. $100-130 each with the exception of the moon/sun rattle, $300-325.

Seven double-sided rattles, unmarked (other side).

Four double-sided round rattles, all made in Japan (one side). $125-150 each.

Four double-sided round rattles, all made in Japan (other side).

Trio of baby rattles with faces, the most unusual of which has a double handle and a whistle in the top of the head, all made in Japan. The more elaborate model with the whistle: $230-250. The other two: $185-200 each.

Two rattles, each with an inserted Dutch girl, one in yellow and one in pink. Each includes paper labels reading "Made in Japan." $265-290 each.

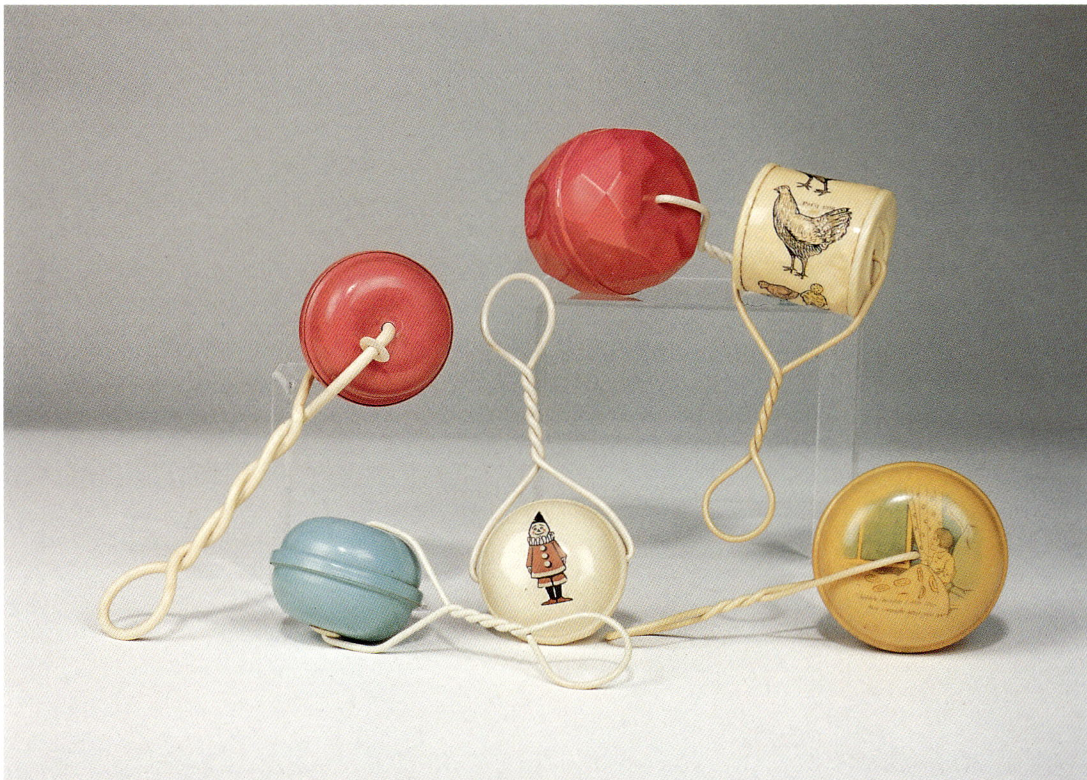

Six rattles, all of which turn—the most unusual rattles are "Twinkle, twinkle little star" and the farm animal painted rattle (patented 1916). Simple to more elaborate: $85-150 each. Early handles and picture rattles, $185-215.

Four egg-shaped rattles, the most unusual of which has a clear plastic top through which a celluloid rabbit may be seen, all unmarked. Simple to more elaborate: $85-100 each.

Two rattles, bull dog and bell, both of which are "Irwin" marked and are of later cellulose acetate composition. $85-100 each.

Two rattles, clear with pink and red backs, both of which have paper circles inside for decoration, American made. $95-110 each.

Two unusual rattles (or possibly teething rings), one with moving eyes and one cat in ring "krazy kat," unmarked. $265-300 each.

Trio of round drum-type rattles, with embossed and attached celluloid figures, unmarked. Prices rise with increasing adornment: $95-165 each.

Two jointed rattles, with painted details, both unmarked and of later vintage. $100-115 each.

Unusual rattle with metal clangers on two celluloid extensions of rattle. Unmarked, but of early 1920s vintage. $100-125.

Four crib toys employing wooden beads on elastic strings, all made in Japan. $135-150 each.

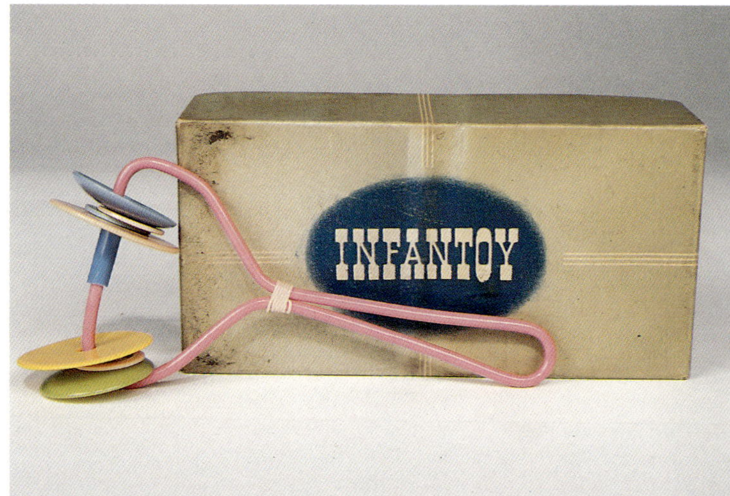

Rattle in original box produced by INFANTOY of New York City. $45-55.

Two crib toys, one with wells at bottom of feet, both completely made from celluloid and made in Japan. $50-75 each.

Cream colored cellulose acetate baby in blue stroller with original box, "A Jeryco Product," made in USA. $75-90.

This elaborate playset for a little girl includes a powder container, comb, and brush, American made. $65-75.

Brush and comb with white ribbons, decorated with pink and blue flowers, in original box, American made. $65-75.

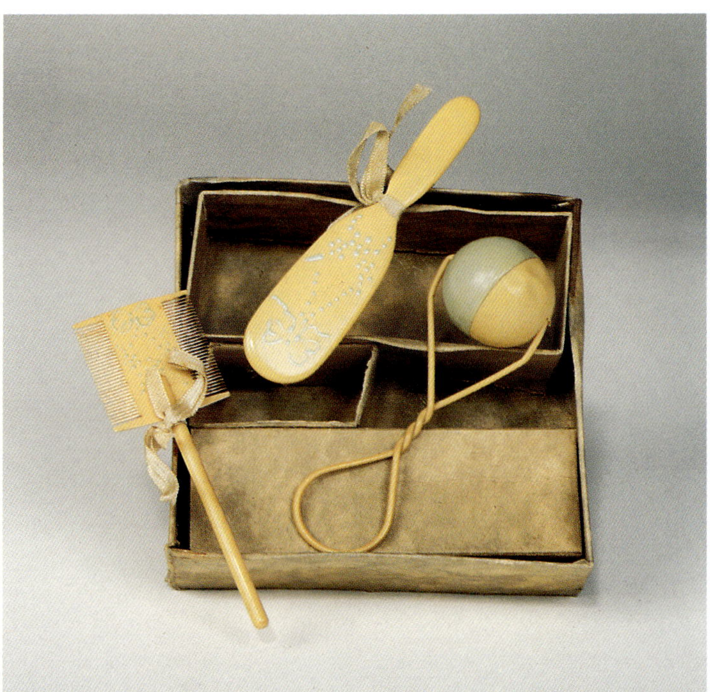

Double edged comb, brush, and tiny roll (yellow and blue) rattle in original box, American. $75-85.

Baby's Christmas present with teething ring and garters with blue ribbon and pink flowers in colorful, lithographed box, American manufactured. $65-75.

Children's spoon and fork with bear figures on handle tops, marked "Athoware, U.S.A. Design Pat. Reg." $130-150 for set.

Miniature pink buggy with tiny celluloid-headed baby, unmarked. $40-50.

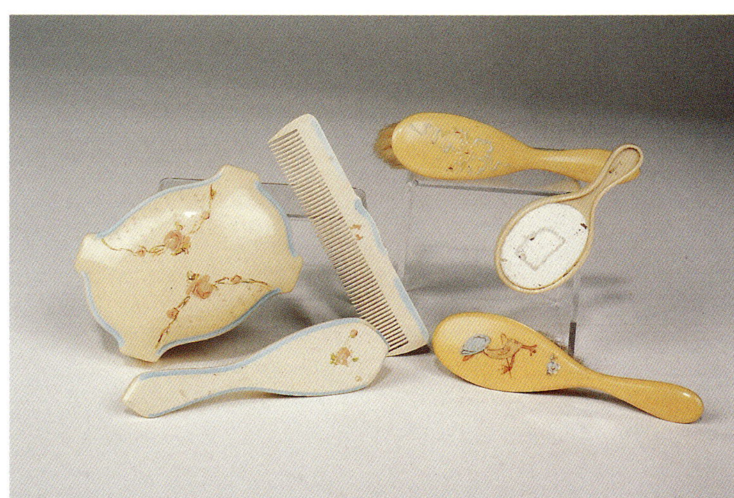

Three brushes, one comb, soap container, and mirror—all simple, yet beautiful children's personal grooming and hygiene items. $50-65 for each set.

Chapter 5.
Novelty and Other Miscellanies

Certain items are just difficult to categorize. These difficult items are placed in this category. Novelty items are often objects that were either purchased while people were on vacation, purchased as little gifts for friends or relatives, or were inexpensive items purchased from a mail order catalog. When received as gifts, novelties sometimes are treasured and displayed proudly in the home. At other times, these novelties are displayed only when the gift giver comes to visit us. Occasionally, these novelties are discarded and forgotten. We even plead "ignorance" if asked about them in later years. But many of these items are "works of art" which today are recognized by collectors as quite valuable.

A donkey and elephant are mounted on a celluloid platform. This is very possibly an item produced by the Republicans during a convention or election time. The donkey is stubborn and refuses to move (while being attached to a chain) while elephant is up and roaring. Japan. $75-85.

Standing cat on top of a ball in a basket. The purpose of this piece is unknown, it may be merely a decorative table item. Japan. $65-75.

This scotty dog has a tiny perfume bottle attached to its neck. Most likely, this was a souvenir item meant for either a woman or a little girl. Japan. $45-60.

Uncle Sam with his top hat, standing on a cylindrical column. A patriotic, souvenir piece. Japan. $120-135.

An elephant is mounted at the top of layered celluloid steps. Possibly a political promotional piece. $75-90.

Trio of birdcages complete with birds perched on celluloid rings, unmarked. $65-75 each.

Twin celluloid birds in celluloid cage and decorated rings below finished with strings, Japan. $75-85.

Celluloid bird perched in a ring incorporated as part of a decorative bird cage with wire rings connected by stings with a wooden top and bottom, Japan. $65-75.

Dogs making friends under a red celluloid umbrella, hard base and dogs, Japan. $100-125.

Twin dogs with top legs jointed, celluloid umbrella, made in Japan. $95-105.

Pen holder, two children holding an umbrella, weighted bottom, made in Japan. $125-135.

Set of unmarked chattering teeth. Most likely an item sold under the category of "magic." $65-75.

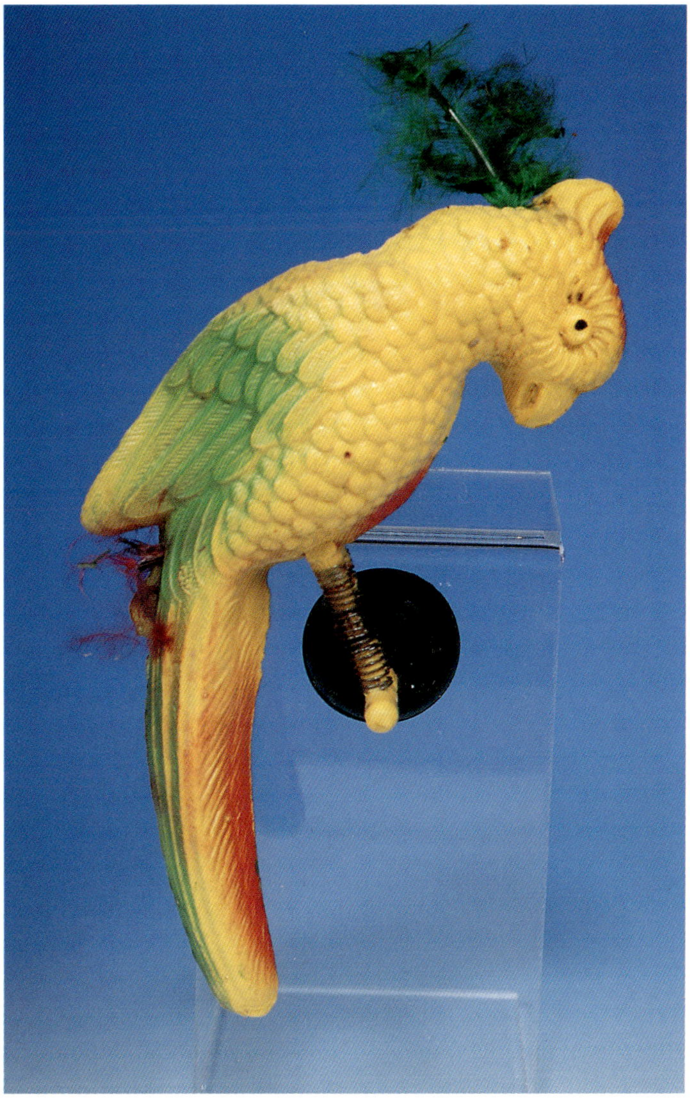

"Irwin" parrot of later cellulose acetate composition with suction cup for attachment, marked "non-inflam." Often these were used as party favors on drink glasses and then recycled somewhere in the home. $45-60.

Original box of party favors, "Made in Japan." Could have doubled as place setting markers at a dinner party as well. $85-100 for set.

Unmarked holder for deck of cards, movable tail points to different suits. $85-100.

Autograph book with a celluloid top attached to the book cover, dated "December 24, 1908. Extremely popular at this time, children, and even adults, would obtain the signatures, little drawings, and even little poems from their friends. $120-135.

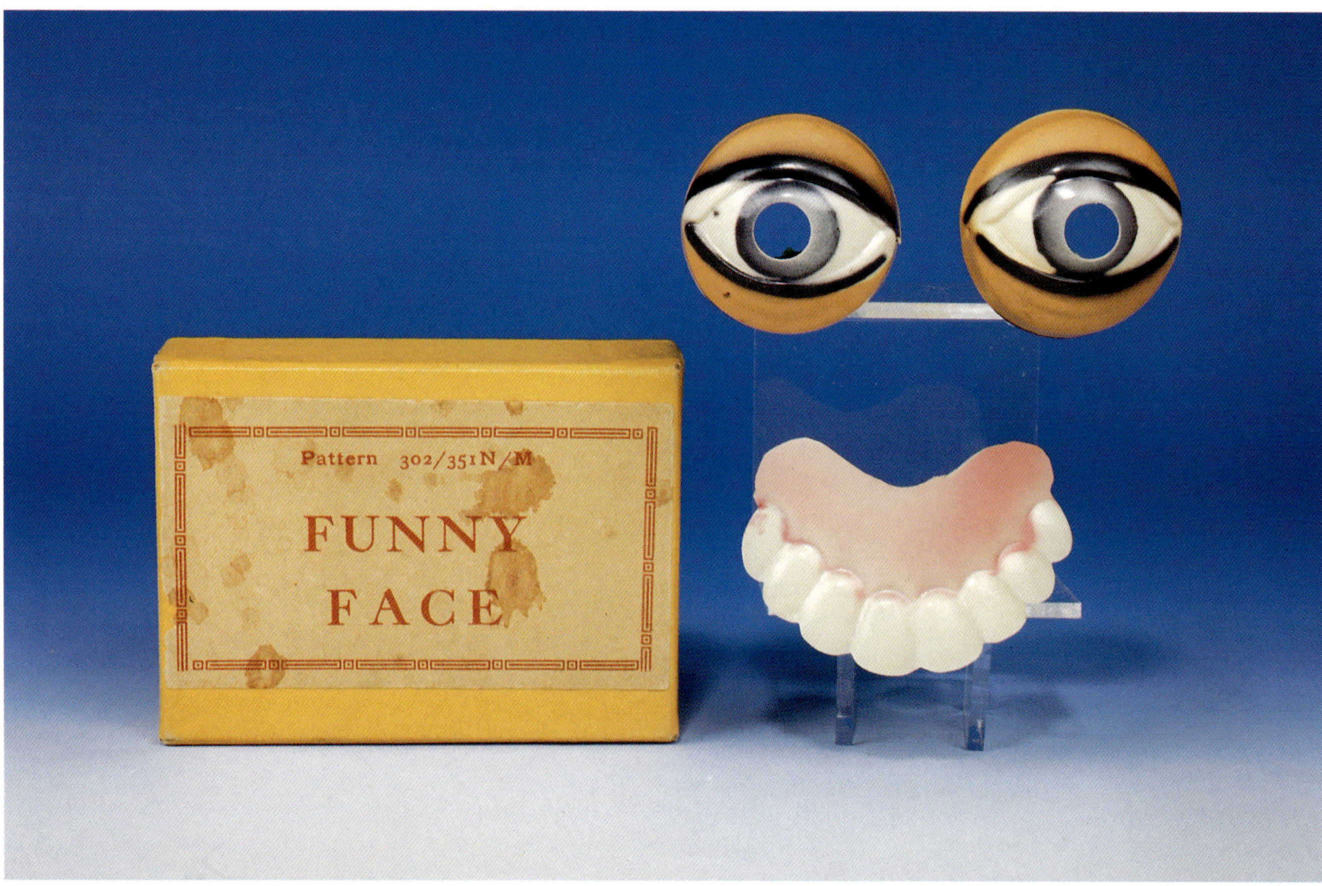

"Funny Face" eyes and teeth in the original box, German. Meant to be used for costume parties and Halloween dress-up. $125-135.

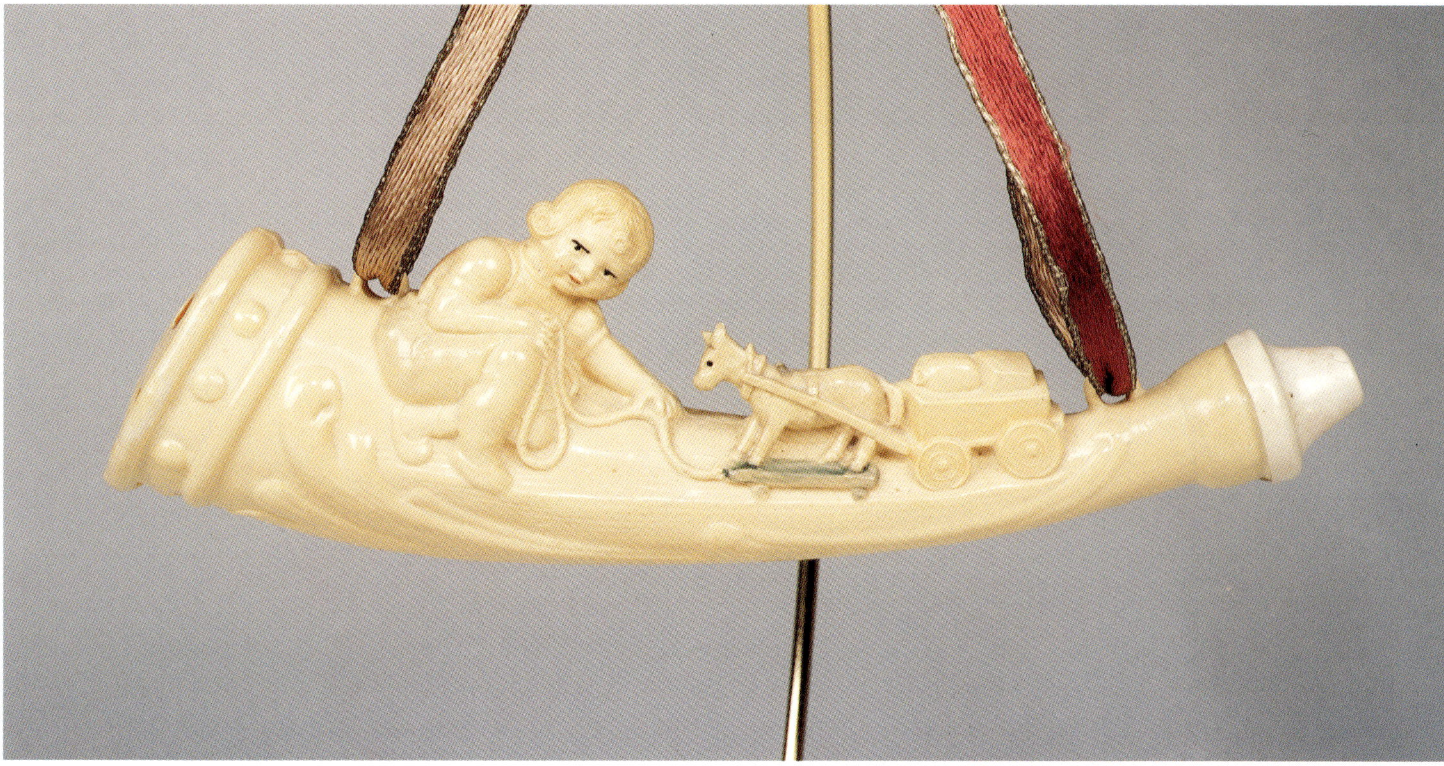

A whistle was inserted into this children's horn with the ribbon attached which features a girl pulling donkey and cart. This whistle has an early HP German marking. Heavier celluloid with an inordinate amount of detail in molding. $400-450.

Back side of the "girl pulling donkey" whistle.

Whistle insert into a children's horn with a ribbon attached. A boy rides a dolphin with open mouth, no markings but detail of molding and heavier celluloid composition point to an early German piece. $395-420.

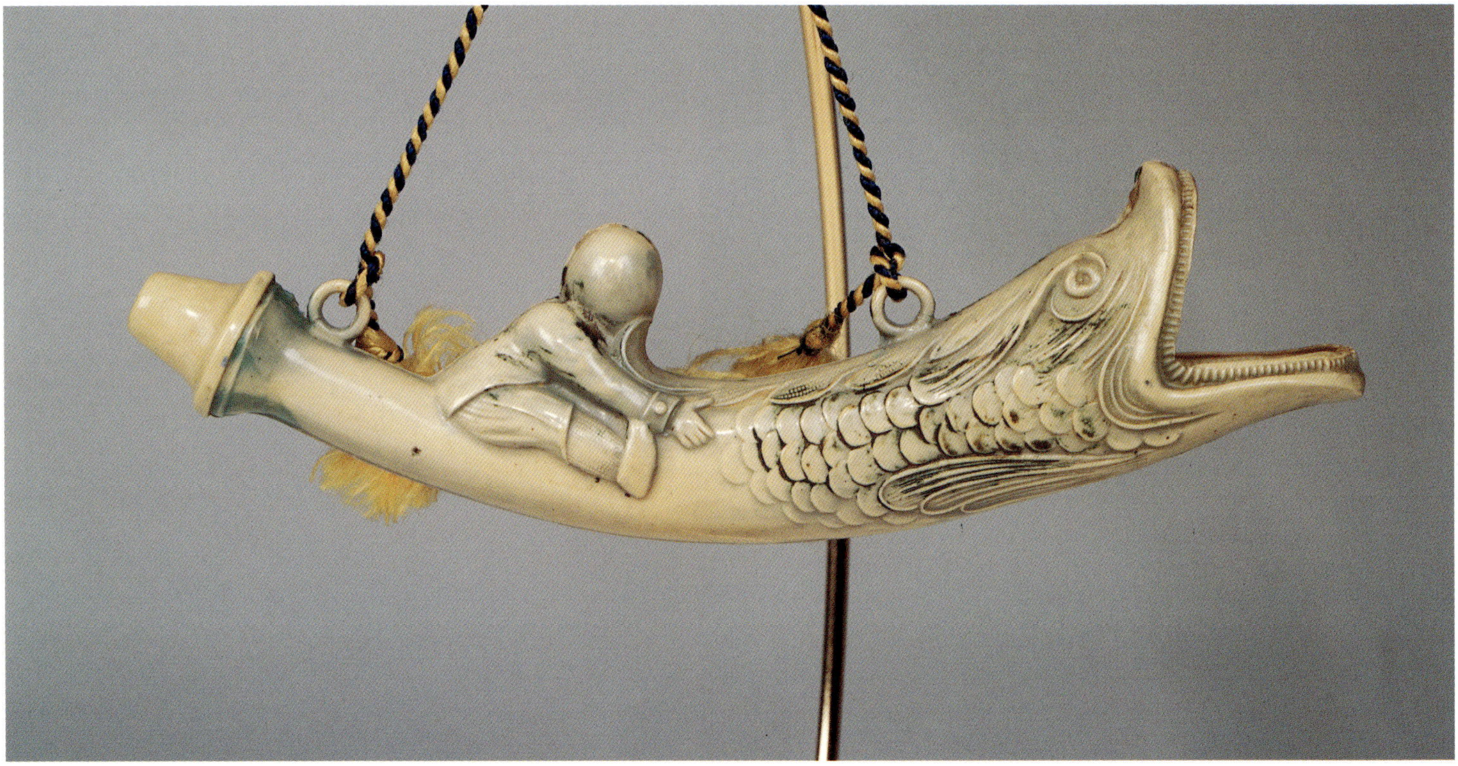

Back of the boy riding the dolphin whistle.

Two "Viscoloid" whistles, red bird and corresponding yellow bird. (Maxine Caldwell collection). $75-85 each.

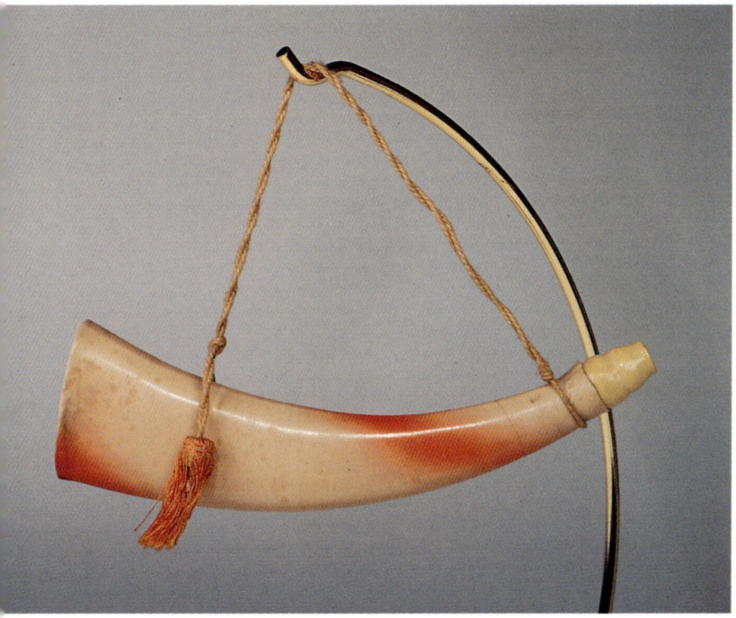

Whistle insert in simple horn, Japanese paper label inside $55-65.

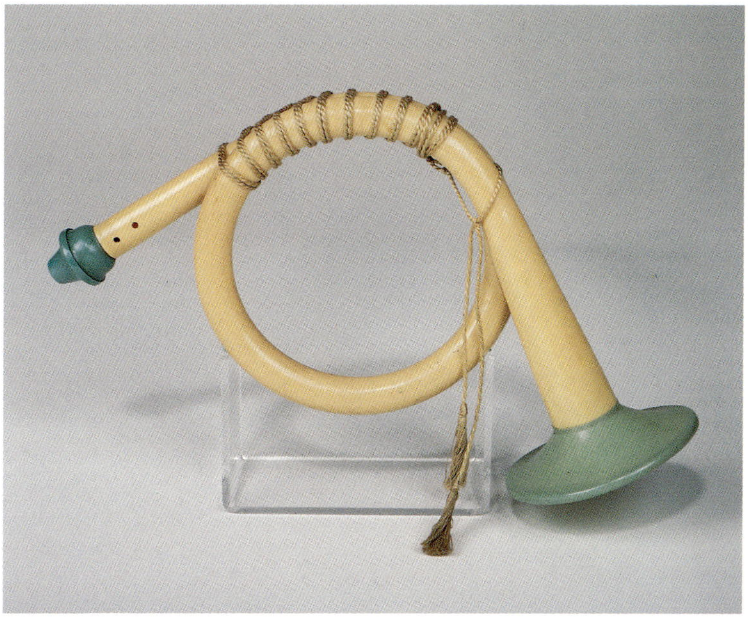

Whistle French horn in cream and blue colors, with pale blue string, American. $75-85.

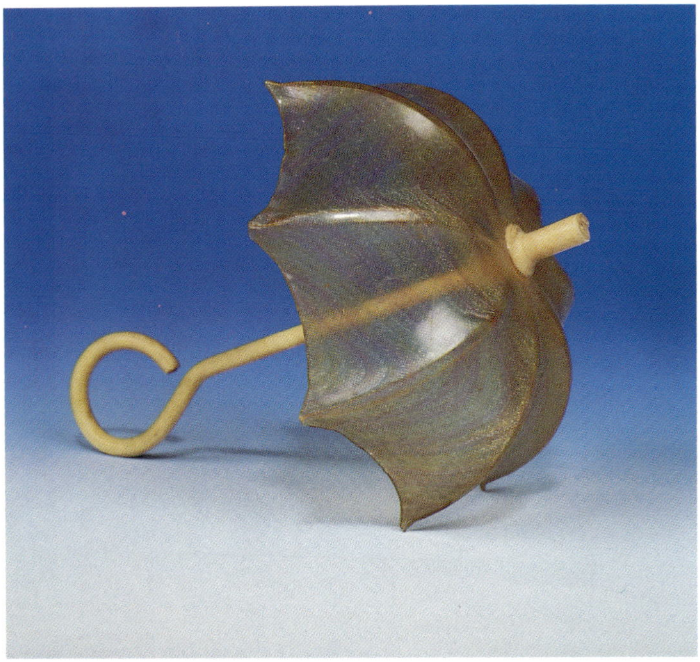

Celluloid umbrella, ivory tones, unmarked. Could have possibly been a children's play toy, but the faint stamping points to a possible souvenir piece. $45-55.

Penis and testicles roly poly. One of the more unusual subjects among the novelty items. One can only conjecture its original selling purpose. Likely made in England. $155-175.

Cards and Greetings on Different Holidays

The early easel Christmas, Valentines, and Easter cards are often among the most elegant of greeting cards created for Americans by both the Germans and by other American lithographers. Many of these cards employed celluloid as a decoration.

Christmas card (5" x 7"), hand-painted celluloid combined with a pulled blue silk fabric on cardboard, American-made with inserted German lithographed card with verse. $75-100.

Mechanical postcard with a silk backing upon which an easel-backed pelican drawn cart of flowers is pulled with a dove carrying a blue flower perched above. Made in Germany. $135-145.

Mechanical postcard with silk trim upon which a celluloid deer-drawn car filled with embossed silk flowers is pulled. Dated December 25, 1909, this card was made in Germany. $125-140.

Valentine card with riveted celluloid and parchment with ribbons from the early 1900s. Easel-backed. American-manufactured. $100-125.

McLoughlin produced Valentine with celluloid flaps from the early 1900s. American-manufactured. $40-50

Satin pillow on a postcard to which a celluloid "A Happy Birthday" message with a basket of flowers has been attached, marked "TRG" Germany. $55-65.

McLoughlin produced Valentine with a riveted red celluloid heart. Meant to hang on a wall. $145-165.

Advertising

Since celluloid was inexpensive to produce, countless advertising items were manufactured during its history. Even commemorative and souvenir items abound. Common objects include bookmarks, advertising calendars, matchbox and pill holders, blotters, letter openers, and trays. Pocket mirrors, buttons, combs, measuring devises, and even beer foam skimmers also were produced in huge quantities.

Celluloid coated lithographed cardboard sign for "International Grape-Cream Baking Powder" for International Food Co. Minneapolis, Minn. USA, American made. $335-450.

Five by seven-inch "Holiday Umbrellas" celluloid advertising sign, unmarked. $165-190.

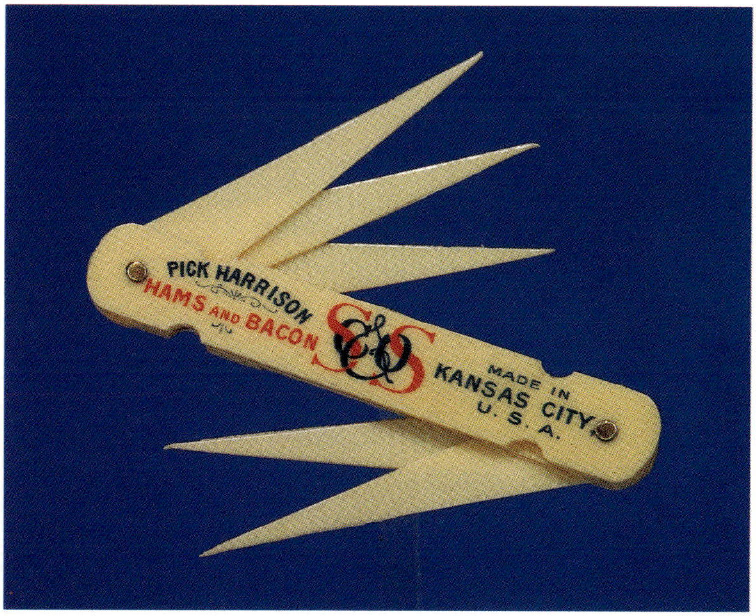

Miniature advertising jackknife. "Pick Harrison from Kansas City" on one side and "Schwarzschild & Sulzberger Co." American made (Maxine Caldwell collection). $125-140.

Advertising for "Riley Oil Co. of Eagleville, Mo." celluloid frame with thermometer and glass insert displaying birds in background, American made. $95-105.

Advertising blotter with a celluloid pin and cover attached, dated January 1, 1915. This item was a give-a-away from the Inter-State National Bank, Kansas City (patent date for blotter is June 8, 1908). $75-90.

Advertising card with an archer, a frog, and an apple entitled "Wear Celluloid Collars and Cuffs." $65-75.

Two small trade cards advertising Celluloid collars, cuffs, and shirt buttons. $35-45 each.

Two advertising cards— a salesman selling products to Orientals doing laundry and an Oriental girl advertising celluloid clothing accessories, both from Donaldson Brothers, Five Points, New York. $40-50 each.

Advertising mirror with a collapsible handle, from "Delaporte's Department Store, Green Bay, Wisconsin," American-made. $165-180.

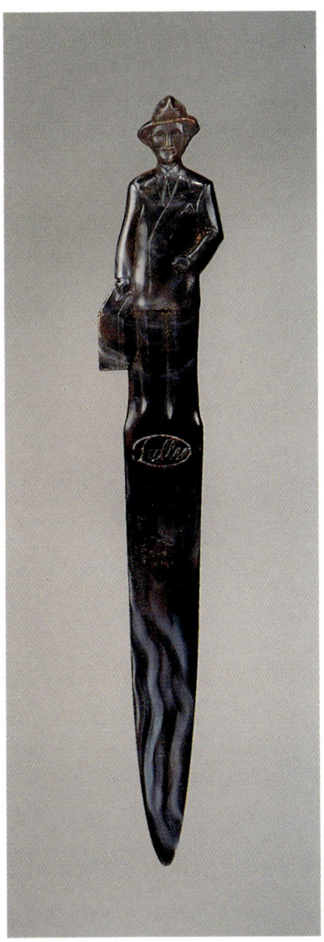

A letter opener given as a premium by the Fuller Brush salesman with a suitcase as a top decoration. $35-45.

Moxie advertising fan reading "Drink Moxie." American manufactured. $185-200.

Tape measures advertising various products "Sucaryl and Enfamil," made in Japan and "Mead Johnson," made in USA. These appear to have been made of the later cellulose acetate composition. $75-100 each.

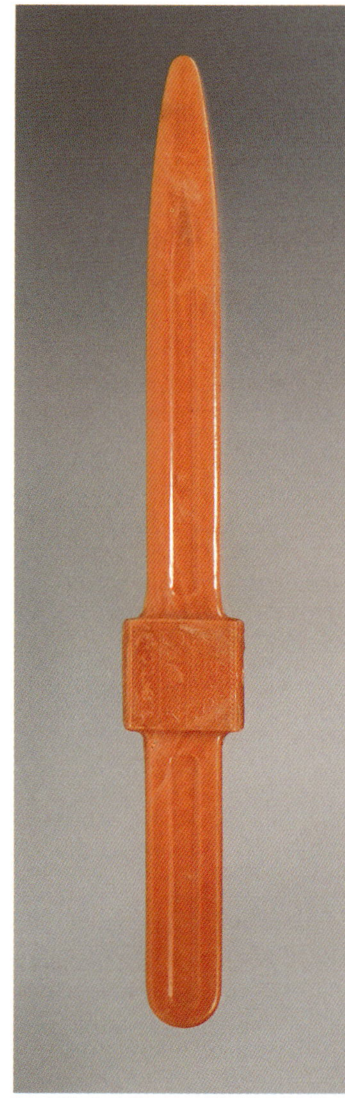

Orange colored letter opener advertising "Macklanburg Duncan Co. Oklahoma City 'Quality Products'." $45-55.

Bear needle holder advertising "American Independent Gypsum Co., Fort Dodge, Iowa 'Up to Minute Plaster Maker'" Patented June 6, 1908. $150-165.

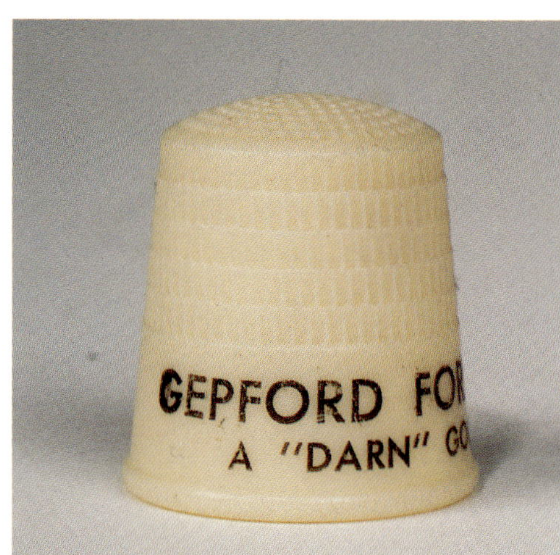

Political advertising thimble, "Gepford for Prosecutor, A 'Darn' Good Democrat." $50-65.

Jewelry

Despite the popularity of celluloid combs, vanity sets, and other such utilitarian objects, its interest among consumers as jewelry was slow to grow. At first most celluloid jewelry was purchased by working-class women, and was largely ignored by the wealthy, who preferred the traditional metal and glass jewelry so long in vogue.

It was not until 1925, when the French designer Coco Chanel introduced the radical notion of costume jewelry, that American women began to take celluloid jewelry seriously (and other such "faux" items) as being worthy of purchase. Chanel produced glass and metal replicas of her own "precious" jewels which started American women in the direction of investigating costume jewelry as an adornment alternative. The popularity of "plastic" jewelry rose sharply with the approach of the Depression years. Very affordable, women soon purchased such jewelry rather than go without such wardrobe accessory items.

One of the first successful producers of celluloid jewelry was George F. Berkander of Providence, Rhode Island. Originally employed in making hair ornaments and barrettes, and impressed with the ivory-like celluloid vanity sets, Berkander experimented with small bar pins made from celluloid. By 1931, he was producing over a thousand gross of inexpensive bracelets a day. His ingenuity also led to his manufacturing of elephant chains for the wives of Presidents Harding and Coolidge. He produced tiny airplanes with Lindbergh's name on them as well. These small planes were ready for sale when the flyer first landed in Paris in May 1927. He also produced the first cellulose acetate flower pins, which soon found their way to chic New York City jewelry stores.

Celluloid necklace simulating ivory, 22-inches. American-produced. $100-125.

Celluloid pendants and pins, all of which simulate the more expensive ivory jewelry of previous decades. Pieces date from the 1920s. Simple to more elaborate: $85-100 each. The prong-shaped item is a personal tooth pick worth $95-100.

"Mary had a little lamb" child's pin, made in Czechoslovakia and seen here on its original card. $65-75.

Variety of Art Deco hat pins in celluloid, the most unusual of which have encrusted paste jewels of various hues. Simple to more elaborate: $20-65 each.

A variety of pins (adult) including peacock, Art Deco, and stylized abstract styles derived from cubist art popular in that time period. Simple to more elaborate (the examples with paste glass jewels): $50-75 each.

Assortment of flower-theme embossed and encrusted pins (adult). Of special note is the center pin with three cut-glass prisms dangling from the pin. Simple to more elaborate: $40-125 each.

Childrens' jewelry, including pins and a dog sweater clasp, made in Japan and the USA. Simple to more elaborate: $45-65 each.

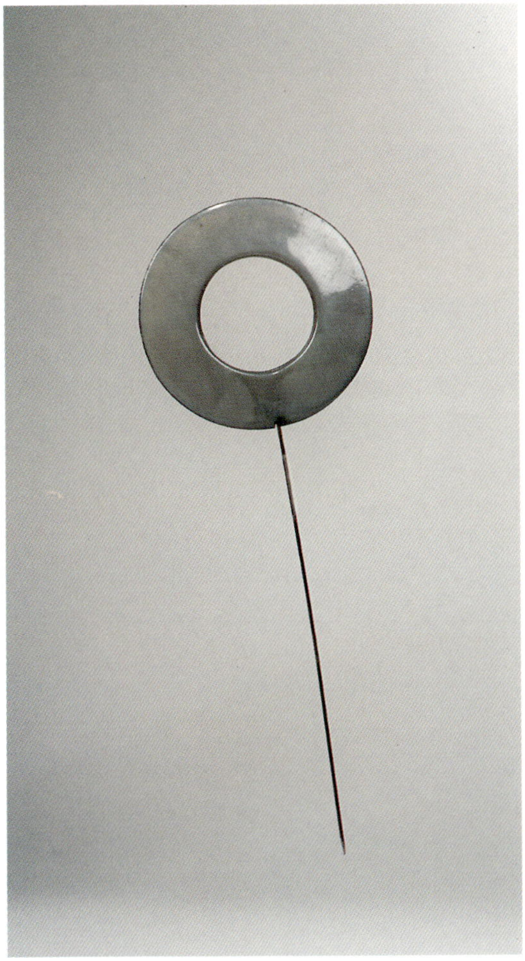

Large seven-inch circular hat pin. American-produced. $25-35.

Mottled green colors popular in the 1920s are reflected in this scarf holder. American-made. $25-35.

Art Deco Pin with stylized lady and Egyptian-theme bracelet, both reflective of this art movement. Pin, $95-105; bracelet, $125-135.

Two simple, yet elegant glass-stone set bracelets meant to simulate more expensive gold and silver jewelry of the period. American-made. $100-125 each.

Pink-toned abstract ring meant to simulate feathers or leaves. American-produced in the early 1920s as it was intended to simulate the more expensive ivory jewelry. $75-85.

Bracelet and ring, both featuring designs drawn from Japanese prints in the 1920s. An attractive flower molding with soft pastel shading was used. Bracelet, $100-125; ring, $65-80.

Two pendants, one with a cameo encircled with paste jewels and the other molded with a grape cluster and green leaves. Right, $45-50; left, $55-60.

Rose-shaped pendant with a necklace of molded celluloid beads, once again meant to simulate ivory. $85-95.

Delicate celluloid beads woven into a lace design on thin wires surrounding pink celluloid roses. Pin, $50-65; earrings, $40-50.

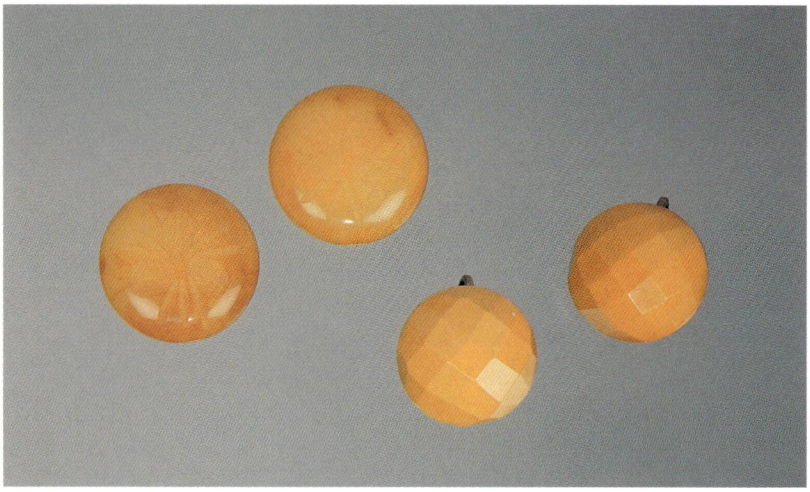

Simple, yet rich looking earrings. Left, $30-40; right, $25-30.

Simulated ivory, button-shaped earrings. $50-65.

Four hair ornaments, the rarest of which have encrusted glass paste jewels, primarily used as ornamentation. Simple to more elegant: $45-65 each.

Three hair combs, the rarest of which once again have encrusted paste jewels, 1915 to 1920. These were meant to be utilitarian, to hold hair in place. Black and amber comb (the rarest of the three), $95-105; tortoise and amber, $75-85; third comb, $25-30.

Chapter 6.

Utilitarian Objects

Since we were looking for inexpensive materials to produce our everyday objects, celluloid began to be employed in the manufacture of countless everyday items including letter openers, knives, dresser sets, picture frames, and even kitchen utensils.

A set of four letter openers: the early molded pieces are in "French ivory." One of the openers is handled with two horses and riders, advertising "Niagara Falls." $60-100, the top left opener is the most expensive.

Set of three letter openers: an Eskimo Indian totem pole, a camel, and a butterfly on a branch. All are unmarked, possibly souvenir items. $35-45 each.

Set of four elaborate letter openers: an elephant ("Viscoloid"), an alligator (German), a mill with a moving wheel (Japan), and an Indian (unmarked). $100-110 each.

Book markers, one red and the other blue, both unmarked. $20-25 each.

A German letter opener given as a Christmas present, shown here in the original stamped box. $85-100.

Blotter with an embossed cover and striated cover sheet over red and green blotting paper, unmarked. $55-65.

Two celluloid book marks, both intricately painted and decorated, unmarked. $40-50 each.

A set of three more elaborate pencil holders: an Indian figure smoking with a pencil pulling from his head feathers, a Japanese lady with a fan from which the pencil is pulled, and an Indian bust with a small pipe pencil inserted in his mouth. Indian, $145-160; Japanese lady, $120-130; Indian bust, $260-275.

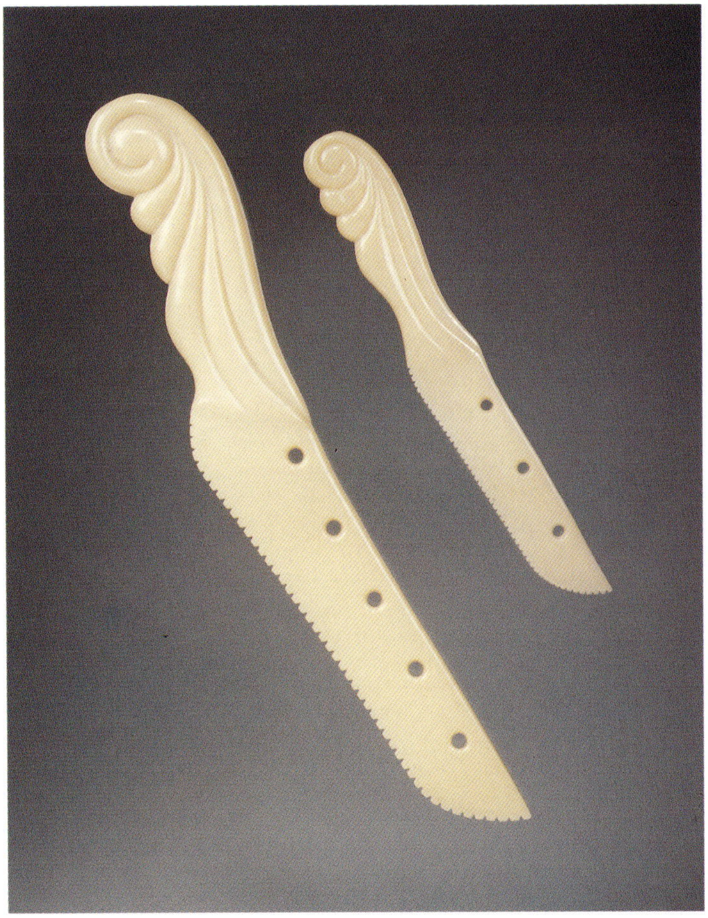

Two serrated knives with holes in the blades, unmarked. $25-35 each.

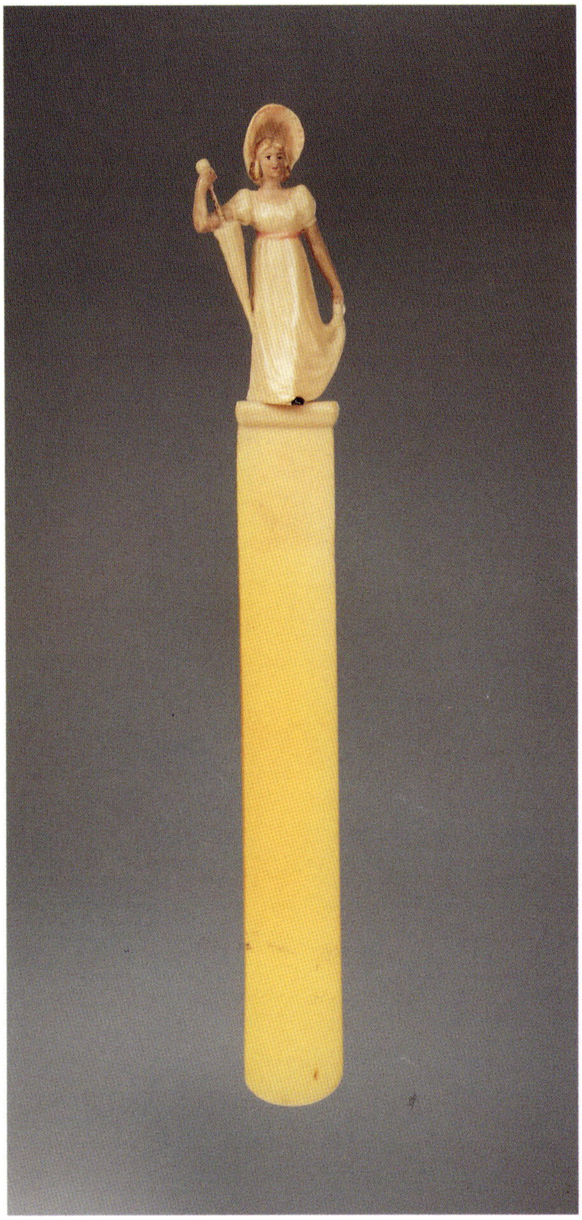

Beer skimmer denoted by its rounded base. Note the chemical changes to the blade. This unmarked skimmer has an elegantly shaped top. $100-110.

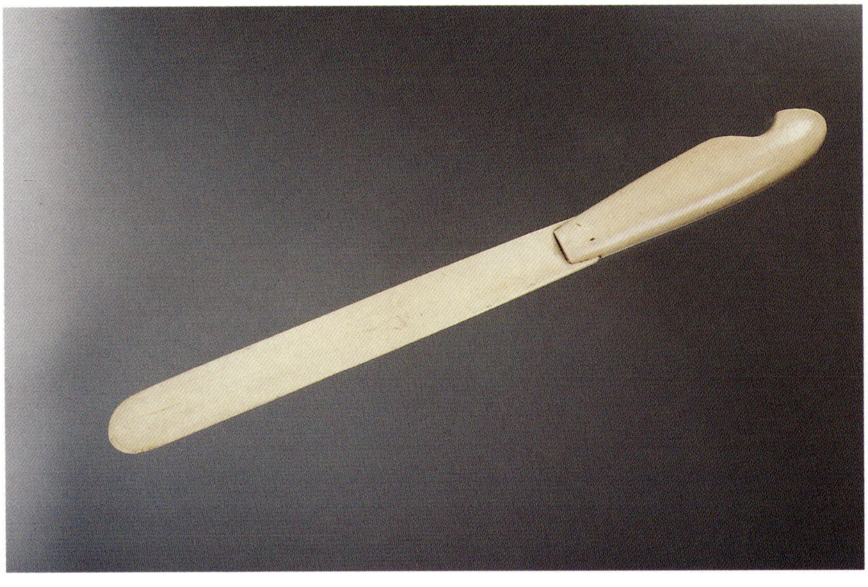

Beer skimmer used to skim foam off glasses of beer, unmarked. $35-55.

Hors d'oeuvres pins, with flower tops, five picks. $5-10 each.

Toothpick dispenser, push down and a toothpick pops up through a hole in the top of the holder, USA. $65-85.

Crumb tray and scraper, marked "Fuller." $35-50.

Three napkin holders, two with names embossed in the celluloid, American. $30-40 each.

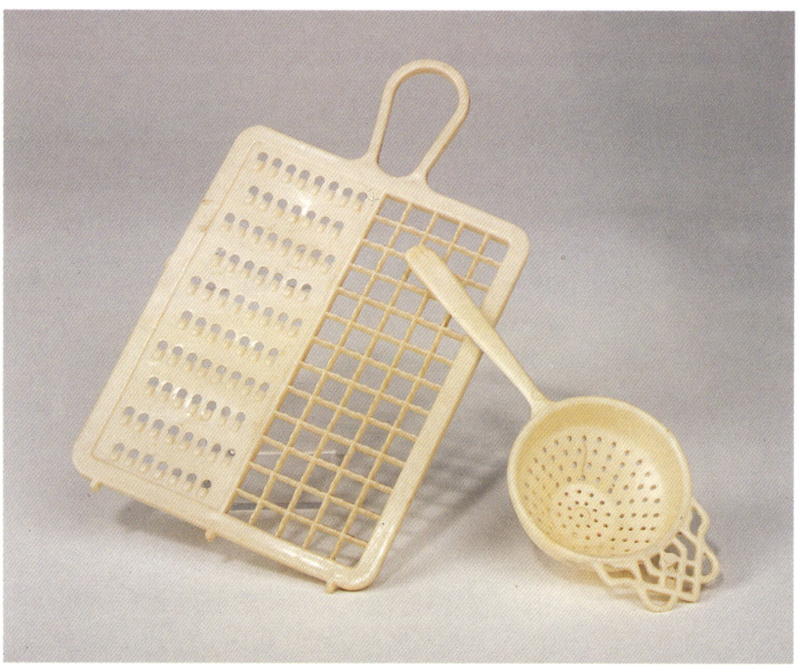

Strainer (marked "Designs, Inc. Hoboken) and grater (marked "Made in USA Patent Pending"). $35-45.

Biscuit and cookie cutter, made in USA, KO brand. $45-55.

Set of bridge score keepers (number 2 marker is missing), with celluloid indicators for number, suit, and no trump, American. Complete set: $65-70.

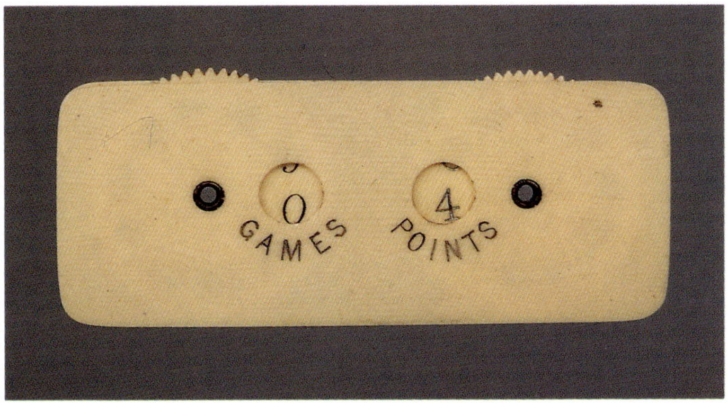

Score keeper for a game, marking game number and points, unmarked. $45-55.

Two sizes of pipes, one still in the original celluloid container (intended as a novelty), American made. The larger of the two is, however, made of Bakelite rather than celluloid. $55-75 each.

Celluloid covered napkin holder with a heart-shaped mirror, Victorian style, American manufactured. $65-75.

A more ornate frame with a metal stand. The celluloid is very 3 dimensionally molded with ornate features. $75-85.

Rectangular frame for a four by six-inch picture, celluloid back, American. $35-40.

Simple oval frame with a celluloid stand, American. $20-25.

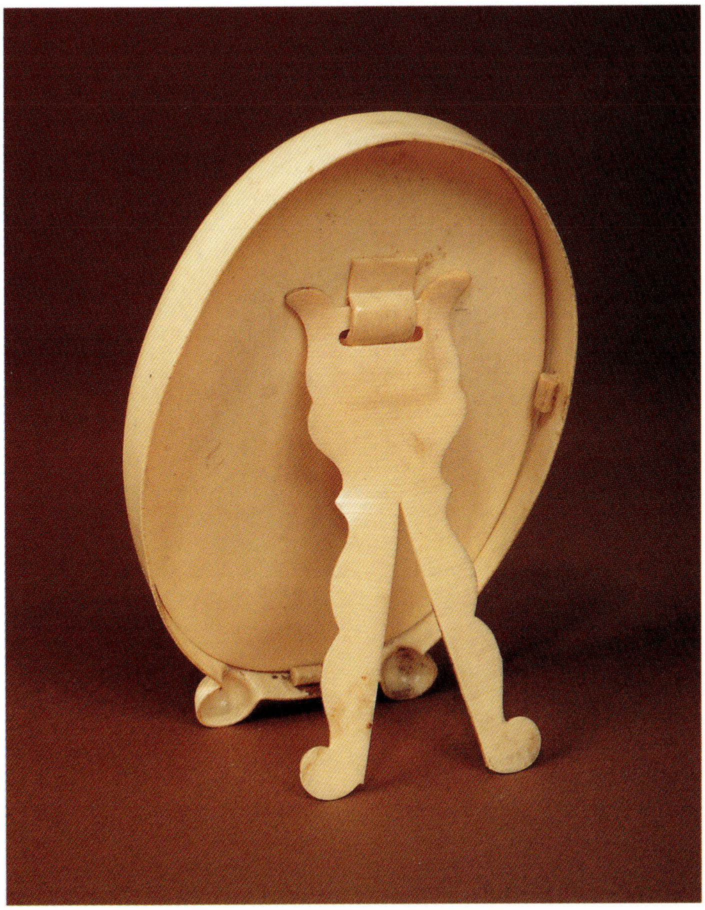

The back of this oval frame illustrates the typical celluloid easel stand.

Marbleized celluloid hanging picture frame, an American-made frame. $65-75.

A set of two frames, one rectangular and the other oval with flowers, American. $35-40 each.

A set of three frames: all are simple renditions of celluloid frames. American manufactured. $25-30 each.

Five celluloid frames: all five are simple and in smaller sizes to be used on table and dresser tops. All five frames are American produced. Smaller to larger: $15-35 each.

Three free standing celluloid frames with easels in back for easy standing on table tops. American made. $45-55 each.

Hanging frame with four ovals. Unusual in size and number of separate photos it could display. American manufactured. $75-85.

Three five by seven inch frames. American made. $35-45 each.

Standing four by six-inch frame. American-made. $30-40.

Double oval (five by seven-inch) frame with a celluloid stand, American. $60-70.

Two frames: the frame on the left is on a celluloid pedestal. American made. $45-65. The frame on the right is larger and has a marbleized-textured celluloid on a celluloid base. $65-75.

Celluloid covered circular picture of a soldier, lithographed on metal, American made. $75-85.

Celluloid covered circular picture of a woman, lithographed on metal, American made. $75-85.

This celluloid picture frame from 1903 was merchandised with a photo taken in a studio and used as a Christmas tree decoration for many years. American manufactured. $125-135.

Celluloid picture with a figure painted on the celluloid. "Germany." $70-80.

Cardboard box with hand-painted and embossed celluloid top, no marking. $75-85.

Cardboard boxes topped with hand-painted celluloid decorative attachments, American manufactured. $65-75 each.

Celluloid topped wicker container for eye glasses, American manufactured. $60-70.

Square green container, brown base, American. $45-55.

Powder puff, top has a butterfly and natural weeds embedded in the celluloid, marbleized base, American made. $95-105.

Collar box, cardboard with celluloid over the entire outer surface, USA. $65-75.

Collar box, opened.

Children's muff with a celluloid face attached to mohair with cloth lining, cloth stuffed feet and hands. Extremely rare and unusual. $425-450.

Two childrens' items — a circular container with clown painted on top and a child's bank with a blue bird on top of the original box, both American. Clown container, $65-75; bank, $55-65.

Wicker woven motif on a container and a pin cushion, both American. Container, $45-55; pin cushion, $30-40.

Turtle pin cushion, made in Japan. $85-90.

Bank, USA, souvenir from the Blackberry Ridge in Spring Green, Wisconsin. $75-85.

Celluloid jewelry box decorated with a celluloid flower, made for the dresser top, USA. $95-105.

One oval and two piano shaped containers, all USA. All of these containers probably held jewelry at some time. Oval container, $30-40; pianos, $75-85.

Three jewelry boxes from E.P. Burman, Jeweler, of 326 E. Commercial Drive, Springfield, MO, American manufactured. $20-35 each.

Jewelry box with a sliding drawer in the bottom, designed to imitate furniture phonographs, USA. $125-135.

Drum table, the top lifts off to be used as a jewelry box. $75-85.

As stated in the history early in this book, most celluloid vanity sets were first produced in France. They were inexpensive reproductions of the more luxurious sets. These became the gift of choice for many children and husbands who purchased such items for birthdays and other holidays.

Most sets contained at least ten pieces, including a comb and brush, at least one hand mirror, a powder box, a hair receiver, small containers for pins and ointments, cylindrical containers for cotton, manicuring implements (scissors and a nail buffer), a tray, an atomizer, and upon occasion, a set would contain a glove stretcher, button hook, and even picture frames.

Their popularity declined as soon as cosmetics were commercially packaged in more elaborate forms. Such elegantly, commercially packaged powders, creams, and lotions could then be proudly displayed on dressers, eliminating the need for the elaborate, formal celluloid dresser sets that had preceded them.

Complete boxed dresser set, in a black carrying case for traveling, American manufactured. $120-130.

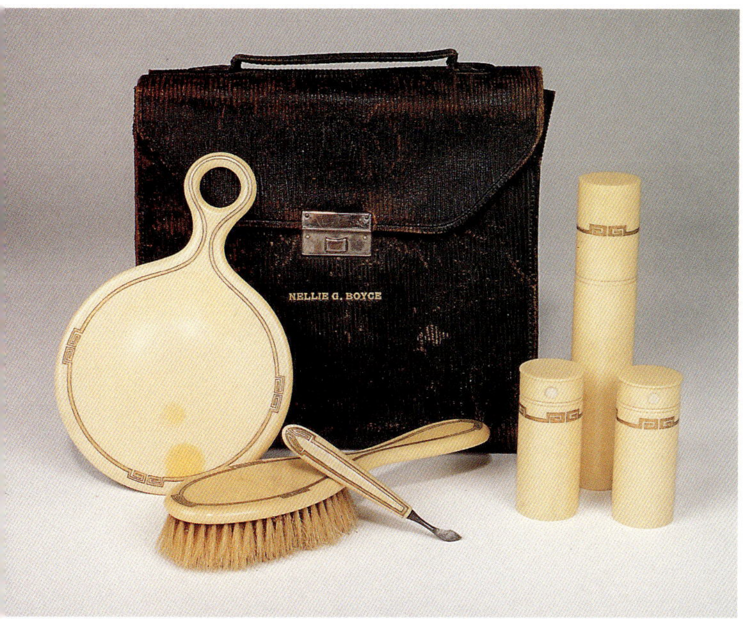

:elluloid dresser set, engraved with the owner's name, "Nellie G. ›oyce." American manufactured. $135-150 complete.

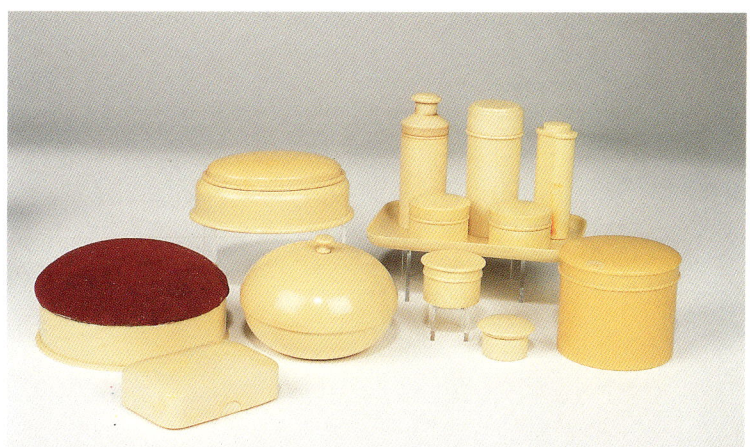

Assortment of dresser items including powder jars, cream jars, trays, perfume dispensers, and other containers, USA. $10-45 each.

Boudoir set with nail files, a button hook for lacing shoes, a powder receptacle, and other containers on a pearlized base. $145-155.

Glass perfume bottle in a celluloid base, American. $65-75.

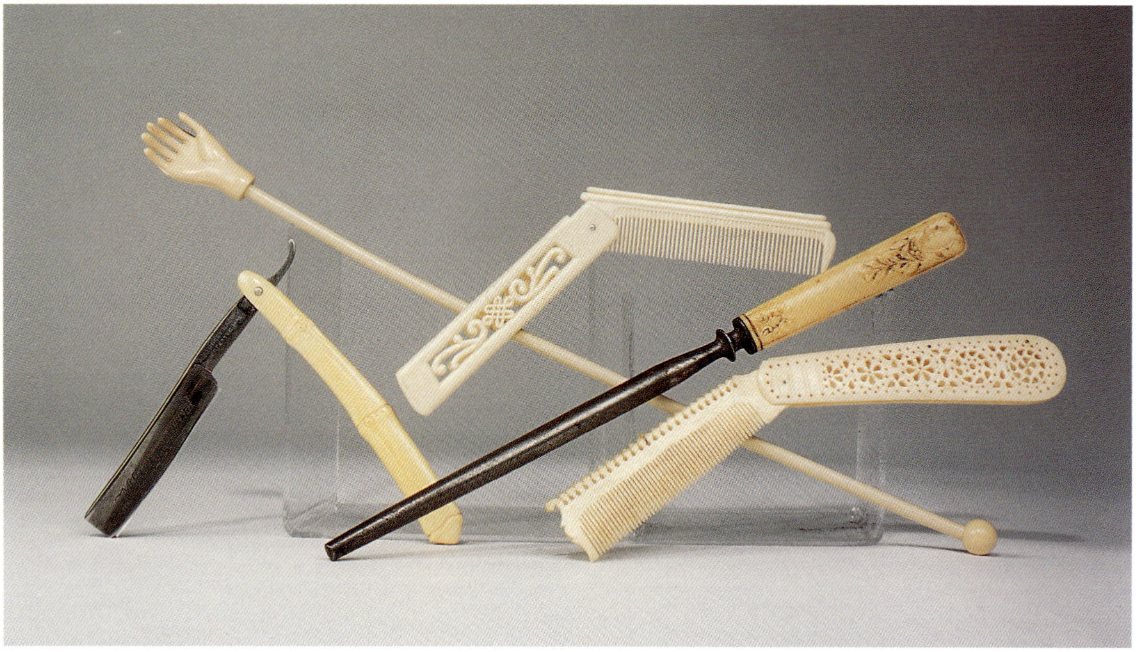

Personal hygiene items including two combs, a hair curler, a straight edge razor, and a back scratcher, all American manufactured. $40-65 each.

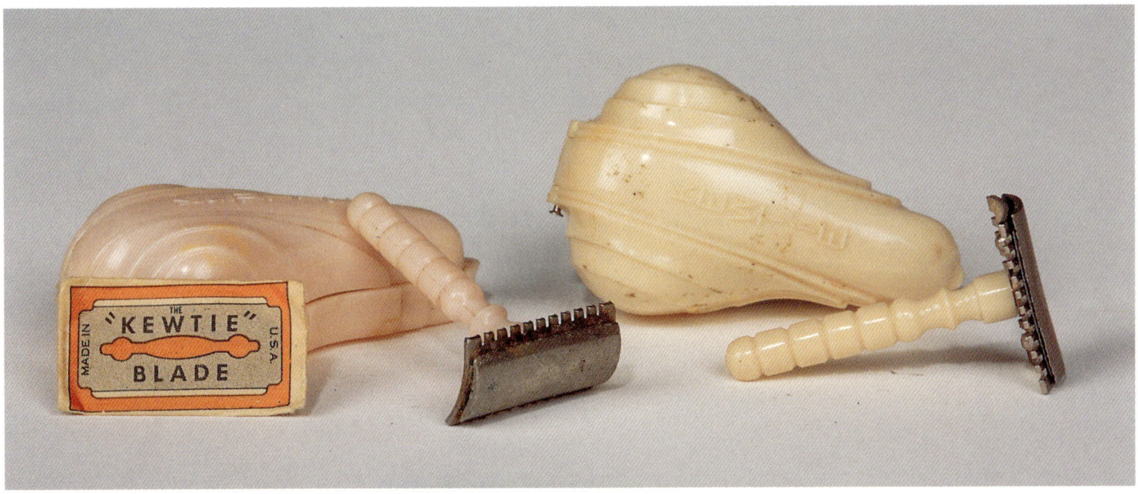

Two miniature ladies' razors, "Kewtie" Blade, U.S.A. These are shown with their pear-shaped celluloid cases celluloid. $40-50 each.

"Viscoloid" toothbrush holder with space for a toothpaste tube at the base. $125-135.

Personal beauty set including a mirror, brushes, a comb, and a cuticle tool, American manufactured. $50-65.

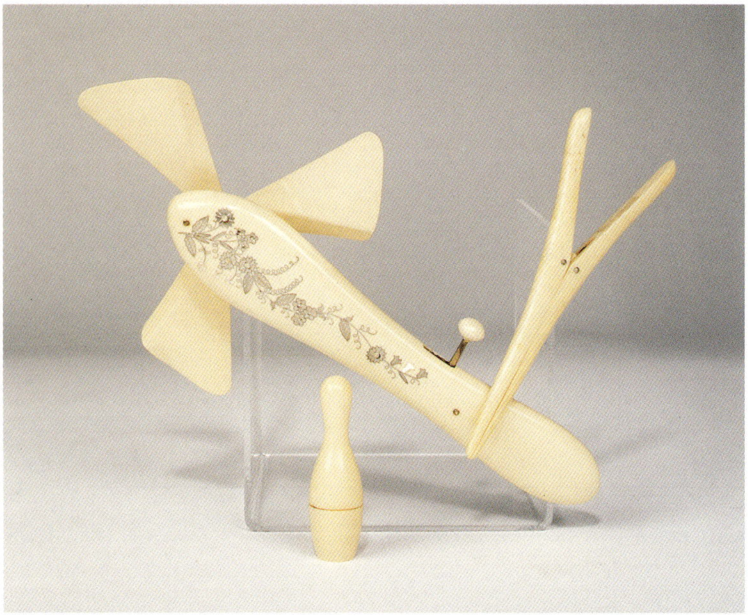

Mechanical opera fan (France), a glove finger stretcher and a bowling pin lighter, USA. Fan, $165-180; glove stretcher, $25-35; bowling pin lighter, $75-85.

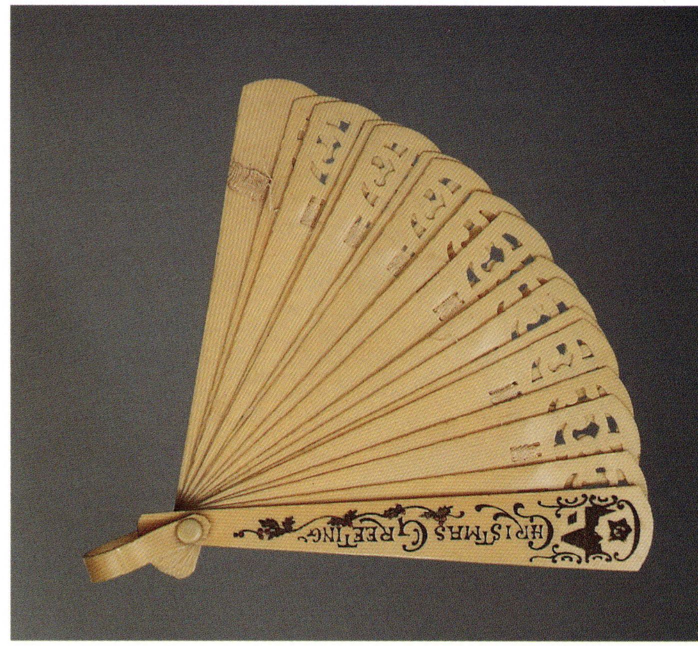

"Christmas Greeting" fan in a simple style, unmarked. $120-130.

Assortment of five fans, in varying sizes and designs, all USA. Small to large: $85-150 each.

Celluloid hair pins with the original container and a hair receiver, USA. $65-75 each. The tortoise piece is $95-105.

Clothes brushes: large, black bristle (1910, made in France) and smaller, white bristle (made in USA) brushes. Small brush, $15-20; larger brush, $25-30.

Opera glasses, American made. $150-160.

A comb is inserted inside this figural piece, a girl with an umbrella held in one hand and a letter in the other. Marked "Germany." $120-130.

Three mirrors: two with stationary handles and the third is a double sided mirror with a folding handle. Left, $35-45; center, $85-95; right, $65-75.

Two-inch cherub holding a mirror with celluloid flowers, "design 20." $85-95.

Cigarette holder with a hat covered head and a cigarette in its mouth, American. $50-60.

Early mirror given by a teacher to her students at Christmas with a celluloid covering, "Miss Lang." $100-110.

Change purses, one with gold trim and a metal clasp, and the other is pink pearlized with celluloid linked carrying strap, along with a pearlized pink fan. American. Left, $60-70; center, $95-100; right, $100-110.

Calendar and a small clock, both American manufactured. Calendar, $65-75; clock, $85-95.

Purse with attached change purse inside, American manufactured. $70-85.

Extremely intricate compact with a hand clasp, the delicate slender fingers of which open and close the mirrored compact. Unmarked. $350-385.

Handkerchief holder, satin lining with ribbon, hand painted outside, American manufactured. $65-75.

Celluloid dresser lamps, celluloid bases with celluloid shades, USA. $250-275 each.

Five vases, two of which have silver colored bases, the tallest of which has a weighted base, USA. Simple to more elaborate: $45-95.

Animal and bird-theme tape measures, all made in Japan, the newest measure is the yellow bird with the pink beak on the far left. $140-160 each.

Two Swordfish tape measures, four-inches in length, and an alligator tape measure, six-inches, Japan. $135-150 each.

Three ship tape measures, Japan. $145-60 each.

Indian, Policeman, market lady, and pirate tape measures, all manufactured in Japan. $165-80 each.

Three baskets of fruit, all three of which are tape measures, Japan. $145-60 each.

Rose-topped tape measure. $90-100.

Two novelty tape measures, one advertising "Lewis Lye" while the other is decorated with roses, both made in Japan. $100-110 each.

A leather trimmed celluloid visors with an identifying tag indicating, "Last worn more than 70 years ago by a clerk at the Topeka Train Depot. This visor was made to last." (Maxine Caldwell collection) $100-125.

Celluloid pool ball, early American manufacture. What a perfect way to end this book on celluloid since a billiard ball was the impetus for the items illustrated within these pages. $100-125.

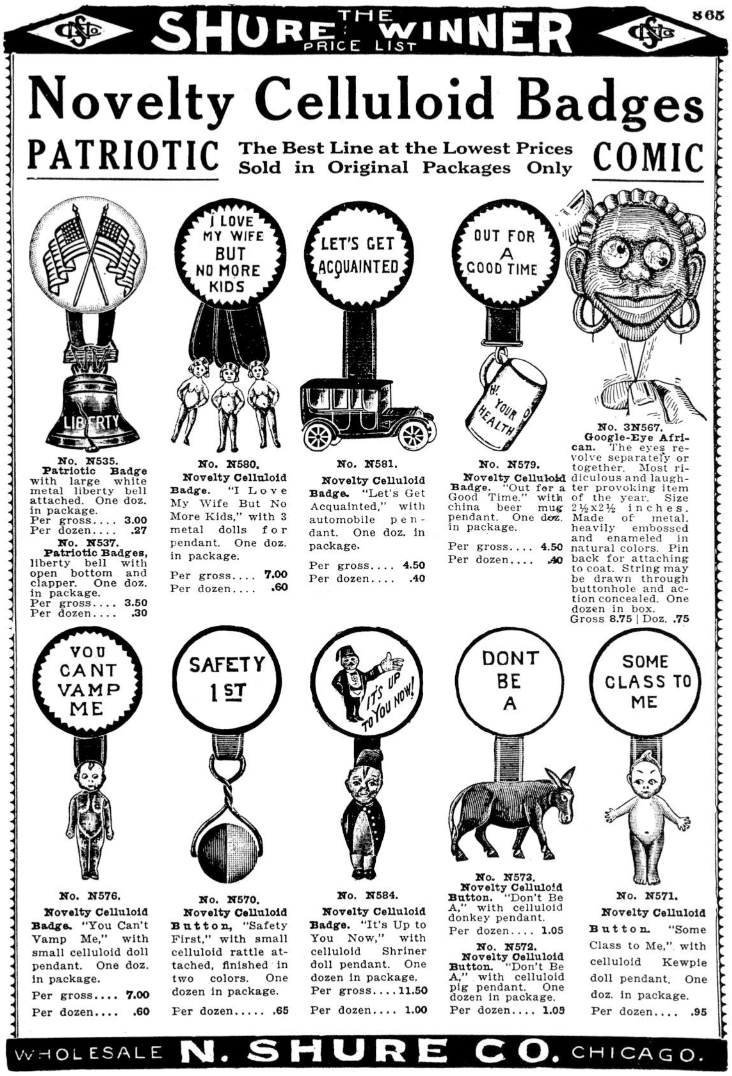
THE SHURE WINNER PRICE LIST 865

Novelty Celluloid Badges

PATRIOTIC The Best Line at the Lowest Prices Sold in Original Packages Only COMIC

No. N535. Patriotic Badge with large white metal liberty bell attached. One doz. in package.
Per gross.... 3.00
Per dozen.... .27

No. N537. Patriotic Badges, liberty bell with open bottom and clapper. One doz. in package.
Per gross.... 3.50
Per dozen.... .30

No. N580. Novelty Celluloid Badge. "I Love My Wife But No More Kids," with 3 metal dolls for pendant. One doz. in package.
Per gross.... 7.00
Per dozen.... .60

No. N581. Novelty Celluloid Badge. "Let's Get Acquainted," with automobile pendant. One doz. in package.
Per gross.... 4.50
Per dozen.... .40

No. N579. Novelty Celluloid Badge. "Out for a Good Time," with china beer mug pendant. One doz. in package.
Per gross.... 4.50
Per dozen.... .40

No. 3N567. Google-Eye African. The eyes revolve separately or together. Most ridiculous and laughter provoking item of the year. Size 2½x2½ inches. Made of metal, heavily embossed and enameled in natural colors. Pin back for attaching to coat. String may be drawn through buttonhole and action concealed. One dozen in box.
Gross 8.75 | Doz. .75

No. N576. Novelty Celluloid Badge. "You Can't Vamp Me," with small celluloid doll pendant. One doz. in package.
Per gross.... 7.00
Per dozen.... .60

No. N570. Novelty Celluloid Button, "Safety First," with small celluloid rattle attached, finished in two colors. One dozen in package.
Per dozen..... .65

No. N584. Novelty Celluloid Badge. "It's Up to You Now," with celluloid Shriner doll pendant. One dozen in package.
Per gross....11.50
Per dozen.... 1.00

No. N573. Novelty Celluloid Button. "Don't Be A," with celluloid donkey pendant.
Per dozen.... 1.05

No. N572. Novelty Celluloid Button. "Don't Be A," with celluloid pig pendant. One dozen in package.
Per dozen.... 1.05

No. N571. Novelty Celluloid Button. "Some Class to Me," with celluloid Kewpie doll pendant. One doz. in package.
Per dozen.... .95

WHOLESALE N. SHURE CO. CHICAGO.

1924

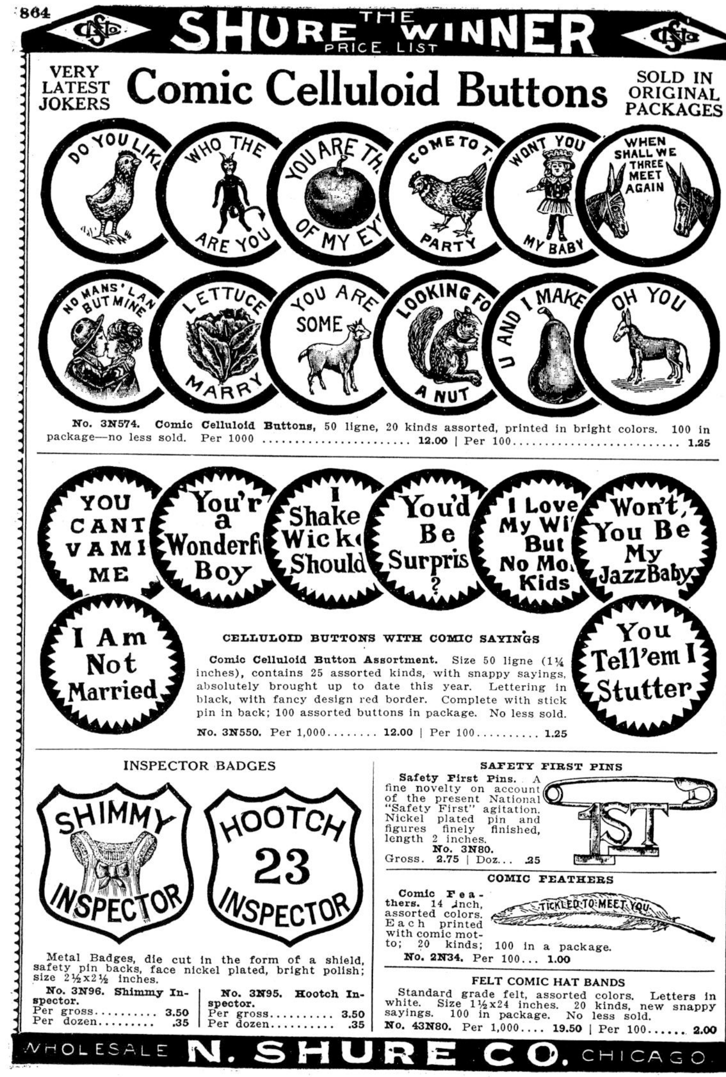
864 THE SHURE WINNER PRICE LIST

VERY LATEST JOKERS

Comic Celluloid Buttons

SOLD IN ORIGINAL PACKAGES

No. 3N574. Comic Celluloid Buttons, 50 ligne, 20 kinds assorted, printed in bright colors. 100 in package—no less sold. Per 1000 .... 12.00 | Per 100 .... 1.25

CELLULOID BUTTONS WITH COMIC SAYINGS

Comic Celluloid Button Assortment. Size 50 ligne (1¼ inches), contains 25 assorted kinds, with snappy sayings, absolutely brought up to date this year. Lettering in black, with fancy design red border. Complete with stick pin in back; 100 assorted buttons in package. No less sold.

No. 3N550. Per 1,000........ 12.00 | Per 100.......... 1.25

INSPECTOR BADGES

Metal Badges, die cut in the form of a shield, safety pin backs, face nickel plated, bright polish; size 2½x2½ inches.

No. 3N96. Shimmy Inspector.
Per gross.......... 3.50
Per dozen......... .35

No. 3N95. Hootch Inspector.
Per gross.......... 3.50
Per dozen.......... .35

SAFETY FIRST PINS

Safety First Pins. A fine novelty on account of the present National "Safety First" agitation. Nickel plated pin and figures finely finished, length 2 inches.
No. 3N80.
Gross. 2.75 | Doz... .25

COMIC FEATHERS

Comic Feathers. 14 inch, assorted colors. Each printed with comic motto; 20 kinds; 100 in a package.
No. 2N34. Per 100... 1.00

FELT COMIC HAT BANDS

Standard grade felt, assorted colors. Letters in white. Size 1½x24 inches. 20 kinds, new snappy sayings. 100 in package. No less sold.
No. 43N80. Per 1,000.... 19.50 | Per 100...... 2.00

WHOLESALE N. SHURE CO. CHICAGO

1924

Balancing Parrots. Very large Parrots, 3½ in. high, made of celluloid. Beak cut to hold place card. Tail weighted and will stand on glass or edge of anything. One dozen in box.

Fancy Tinted

Each Parrot tinted in many natural colors.

No. 32N89.
Per gross....... **4.50**
Per dozen .... **.40**

Solid Colors

Blue, Red, Yellow, Pink or green.

No. 32N90
Per gross **4.50**
Per dozen ...... **.40**

Balancing Birds. All solid colors in yellow, green and blue colors. One dozen of one color in box.
No. 32N6. Per dozen.............................. **.40**

1924

807

THE SHURE WINNER PRICE LIST

Celluloid Toys, Emblems, Etc.

Sold in Original Packages Only

ANIMAL ASSORTMENT

Big Combination, Domestic and Wild Animals. Made of celluloid, beautifully colored to look life-like. Twelve, in partitioned display box. Six kinds—Horse, Dog, Cow, Tiger, Lion and Polar Bear. **No. 32N64.** Per dozen.. .80

Domestic Animals, made of celluloid handsomely tinted in natural colors, perfect in form and feature, 6 styles assorted, horses, pigs, sheep, dogs, donkeys and cows. One dozen assorted in display box; average height 2 inches. **No. 32N42.** Per dozen...... .80

COLORED CELLULOID SWAN

Highly Colored Celluloid Swans. These Swans are offered in two sizes. Both are nicely decorated in many bright colors. Each size, one gross to box.

No. 32N86. Small Swan, 1⅝ inches long. Per gross. 1.50 | **No. 32N87.** Large Swan 2 inches long. Per gross. 2.75

SWANS AND DUCKS

Assorted Swans and Ducks. Made of white celluloid, beautifully embossed and colored. They come assorted, swans and ducks. Swans about 2½x2 inches. Ducks about 2½x1¾ inches. Six dozen to box. **No. 32N88.** Per gross....... 4.50 Per box of 6 dozen.......... 2.40

FLOATING TOYS

Floating Toys, 2½ to 3 inches, assorted ducks, fish, frogs, etc., hollow celluloid bodies, painted in natural colors, 3 kinds. One dozen assorted in box. **No. 32N50.** Per dozen........ .42

FLOATING TOY ASSORTMENT

VISCOLOID

An Assortment of Three Dozen Celluloid Toys that makes a bright spot in the story. Every color. Toys to sell for 5, 10, 15 and 25 cents. Conservative retail value $5.00. Packed in a handsome partitioned display box with sign. **No. 32N25.** Per assortment 3 dozen.................. 3.50

Floating Celluloid Fish, perfect in form, length 4 inches; is an exact reproduction. Comes in bright red color, large natural colored eyes. Two dozen in box.

No. 32N65. Per box of 2 dozen.......... .84

Floating Celluloid Wild Ducks. Length 3¼ in. Richly tinted in assorted natural colors. A most life-like appearance. One dozen on attractive lithographed display card in box. **No. 32N55.**

Per gross.. 9.50 | Per dozen .80

Wild Animals. Made of celluloid, handsomely tinted in natural colors, perfect in form and proportion. Six styles, assorted—Camel, Elephant, Lion, Tiger, Bear and Rhinoceros. One dozen, assorted, in display box. Average height 2 inches. **No. 32N43** Per dozen........ .80

SHRINERS' CAMELS

Shriner's Camels. Well made Celluloid camels in natural colors. Assorted sizes to suit every requirement.

Plain Camel. Camel 1¾x2¾ inches. One dozen in display box. **No. 32N29.** Per gross....... 8.00 Per dozen .... .70

Camels with Ring in Hump. Camel with brass ring in hump for attaching. Badge. Size 1⅞x2¾ inches.

No. 5N30. Per gross........ 7.50 Per dozen .... .65

Camel, 2¼x2¾ inches, with ring in hump for attaching badge. **No. 5N31.** Per gross........ 9.00 Per dozen................. .80

CELLULOID EMBLEMS

Political Party Celluloid Emblems. Elephants and Donkeys in natural colors, average length about 2¼ inches. For use in making political badges for banquets, etc. Also a nice toy.

No. 32N30. Republican Elephant. Per gross. 8.00 Per dozen .70 | **No. 32N28. Democratic Donkey.** Per gross. 8.00 Per dozen. .70

WHOLESALE N. SHURE CO. CHICAGO.

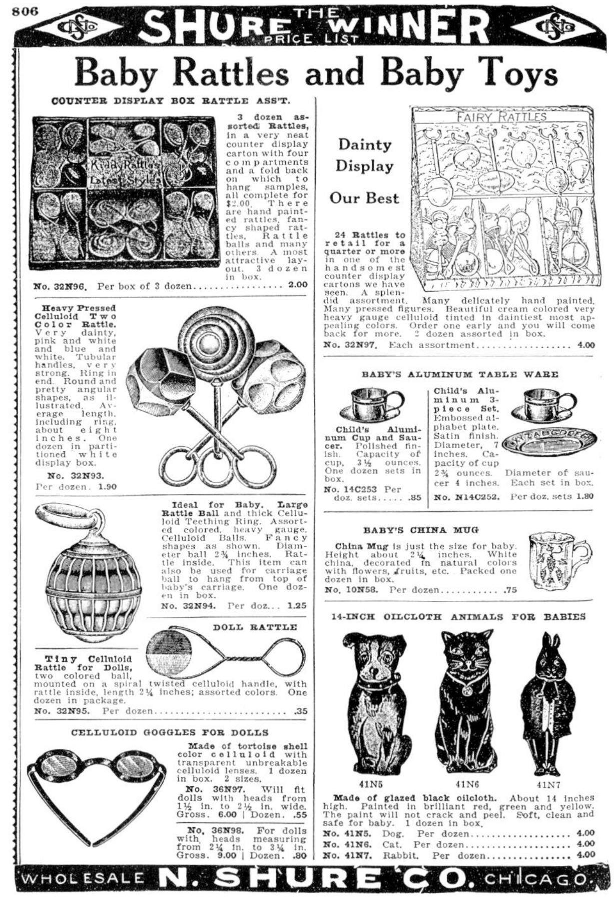
806

THE SHURE WINNER PRICE LIST

Baby Rattles and Baby Toys

COUNTER DISPLAY BOX RATTLE ASS'T.

3 dozen assorted Rattles, in a very neat counter display carton with four compartments and a fold back on which to hang samples, all complete for $2.00. There are hand painted rattles, fancy shaped rattles, Rattle balls and many others. A most attractive layout. 3 dozen in box.

No. 32N96. Per box of 3 dozen................ 2.00

Heavy Pressed Celluloid Two Color Rattle. Very dainty, pink and white and blue and white. Tubular handles, very strong. Ring in end. Round and pretty angular shapes, as illustrated. Average length, including ring, about eight inches. One dozen in partitioned white display box.

No. 32N93. Per dozen. 1.90

Ideal for Baby. Large Rattle Ball and thick Celluloid Teething Ring. Assorted colored, heavy gauge, Celluloid Balls. Fancy shapes as shown. Diameter ball 2¾ inches. Rattle inside. This item can also be used for carriage ball to hang from top of baby's carriage. One dozen in box. **No. 32N94.** Per doz... 1.25

DOLL RATTLE

Tiny Celluloid Rattle for Dolls, two colored ball, mounted on a spiral twisted celluloid handle, with rattle inside, length 2¼ inches; assorted colors. One dozen in package. **No. 32N95.** Per dozen........................ .35

CELLULOID GOGGLES FOR DOLLS

Made of tortoise shell color celluloid with transparent unbreakable celluloid lenses. 1 dozen in box. 2 sizes.

No. 36N97. Will fit dolls with heads from 1½ in. to 2½ in. wide. Gross. 6.00 | Dozen. .55

No. 36N98. For dolls with heads measuring from 2¼ in. to 3¼ in. Gross. 9.00 | Dozen. .80

Dainty Display

Our Best

24 Rattles to retail for a quarter or more in one of the handsomest counter display cartons we have seen. A splendid assortment. Many delicately hand painted. Many pressed figures. Beautiful cream colored very heavy gauge celluloid tinted in daintiest most appealing colors. Order one early and you will come back for more. 2 dozen assorted in box.

No. 32N97. Each assortment................. 4.00

BABY'S ALUMINUM TABLE WARE

Child's Aluminum Cup and Saucer. Polished finish. Capacity of cup, 3½ ounces. One dozen sets in box. **No. 14C253** Per doz. sets..... .85

Child's Aluminum 3-piece Set. Embossed alphabet plate. Satin finish. Diameter, 7 inches. Capacity of cup 2¾ ounces. Diameter of saucer 4 inches. Each set in box. **No. N14C252.** Per doz. sets 1.80

BABY'S CHINA MUG

China Mug is just the size for baby. Height about 2¼ inches. White china, decorated in natural colors with flowers, fruits, etc. Packed one dozen in box. **No. 10N58.** Per dozen........... .75

14-INCH OILCLOTH ANIMALS FOR BABIES

41N5 41N6 41N7

Made of glazed black oilcloth. About 14 inches high. Painted in brilliant red, green and yellow. The paint will not crack and peel. Soft, clean and safe for baby. 1 dozen in box.

No. 41N5. Dog. Per dozen................. 4.00
No. 41N6. Cat. Per dozen................. 4.00
No. 41N7. Rabbit. Per dozen................. 4.00

WHOLESALE N. SHURE CO. CHICAGO.

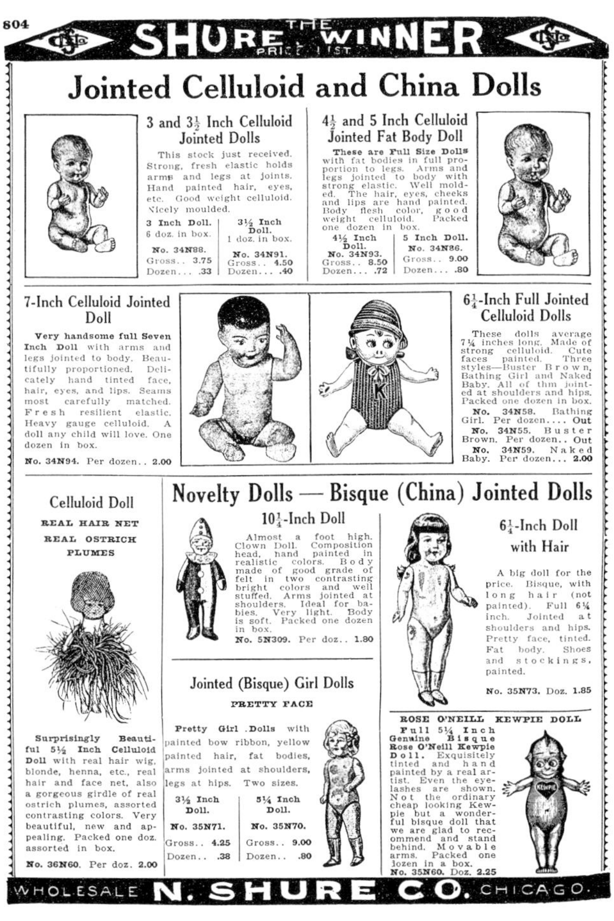
804

THE SHURE WINNER PRICE LIST

Jointed Celluloid and China Dolls

3 and 3½ Inch Celluloid Jointed Dolls

This stock just received. Strong, fresh elastic holds arms and legs at joints. Hand painted hair, eyes, etc. Good weight celluloid. Nicely moulded.

3 Inch Doll. 6 doz. in box. **No. 34N88.** Gross.. 3.75 Dozen... .33 | **3½ Inch Doll.** 1 doz. in box. **No. 34N91.** Gross.. 4.50 Dozen... .40

4½ and 5 Inch Celluloid Jointed Fat Body Doll

These are Full Size Dolls with fat bodies in full proportion to legs. Arms and legs jointed to body with strong elastic. Well molded. The hair, eyes, cheeks and lips are hand painted. Body flesh color, good weight celluloid. Packed one dozen in box.

4½ Inch Doll. No. 34N93. Gross.. 8.50 Dozen... .72 | **5 Inch Doll. No. 34N86.** Gross.. 9.00 Dozen... .80

7-Inch Celluloid Jointed Doll

Very handsome full Seven Inch Doll with arms and legs jointed to body. Beautifully proportioned. Delicately hand tinted face, hair, eyes, and lips. Seams most carefully matched. Fresh resilient elastic. Heavy gauge celluloid. A doll any child will love. One dozen in box.

No. 34N94. Per dozen.. 2.00

6¼-Inch Full Jointed Celluloid Dolls

These dolls average 7¼ inches long. Made of strong celluloid. Cute faces painted. Three styles—Buster Brown, Bathing Girl and Naked Baby. All of thm jointed at shoulders and hips. Packed one dozen in box.

No. 34N58. Bathing Girl. Per dozen.... Out
No. 34N55. Buster Brown. Per dozen.. Out
No. 34N59. Naked Baby. Per dozen... 2.00

Celluloid Doll

REAL HAIR NET

REAL OSTRICH PLUMES

Surprisingly Beautiful 5½ Inch Celluloid Doll with real hair wig, blonde, henna, etc., real hair and face net, also a gorgeous girdle of real ostrich plumes, assorted contrasting colors. Very beautiful, new and appealing. Packed one doz. assorted in box.

No. 36N60. Per doz. 2.00

Novelty Dolls — Bisque (China) Jointed Dolls

10¼-Inch Doll

Almost a foot high. Clown Doll. Composition head, hand painted in realistic colors. Body made of good grade of felt in two contrasting bright colors and well stuffed. Arms jointed at shoulders. Ideal for babies. Very light. Body is soft. Packed one dozen in box.

No. 5N309. Per doz.. 1.80

Jointed (Bisque) Girl Dolls

PRETTY FACE

Pretty Girl Dolls with painted bow ribbon, yellow painted hair, fat bodies, arms jointed at shoulders, legs at hips. Two sizes.

3½ Inch Doll. No. 35N71. Gross.. 4.25 Dozen.. .38 | **5¼ Inch Doll. No. 35N70.** Gross.. 9.00 Dozen.. .80

6¼-Inch Doll with Hair

A big doll for the price. Bisque, with long hair (not painted). Full 6¼ inch. Jointed at shoulders and hips. Pretty face, tinted. Fat body. Shoes and stockings, painted.

No. 35N73. Doz. 1.85

ROSE O'NEILL KEWPIE DOLL

Full 5¼ Inch Genuine Bisque Rose O'Neill Kewpie Doll. Exquisitely tinted and hand painted by a real artist. Even the eyelashes are shown. Not the ordinary cheap looking Kewpie but a wonderful bisque doll that we are glad to recommend and stand behind. Movable arms. Packed one dozen in a box.

No. 35N60. Doz. 2.25

WHOLESALE N. SHURE CO. CHICAGO.

THE SHURE WINNER PRICE LIST

HOOP-LA OUTFIT

Consists of 150 Articles of Merchandise, 50 Six inch Hoop-la Boxes, 100 Six inch Hoops

No. P74

This Hoop-La outfit has been carefully selected and represents the greatest assortment of goods for this purpose. Every article is staple, and useful, and contains such items as French Ivory Clocks, Fine Gold Plated Watches, French Ivory Picture Frames, Fine Pipes and Smokers' Articles, Talcum Powder, Perfume, Flashlights, Cigar Jars, Jewelry, Traveling Sets, Beauty Pins, Etc., Etc. The retail value of these goods ranges up to $2.00.

The goods are very flashy and especially adapted for this business and represent unusual values, both for the operator and the customer. With this outfit you are assured of the big play and enormous profits, and there will be no dull moments once the crowd looks over this assortment.

No. P74—PER ASSORTMENT COMPLETE 30.00

No. P75—Per Assortment of Merchandise Without Boxes and Hoops . . . 20.75

Top Left: 1924
Top Right: 1924
Above: 1924
Left: 1924

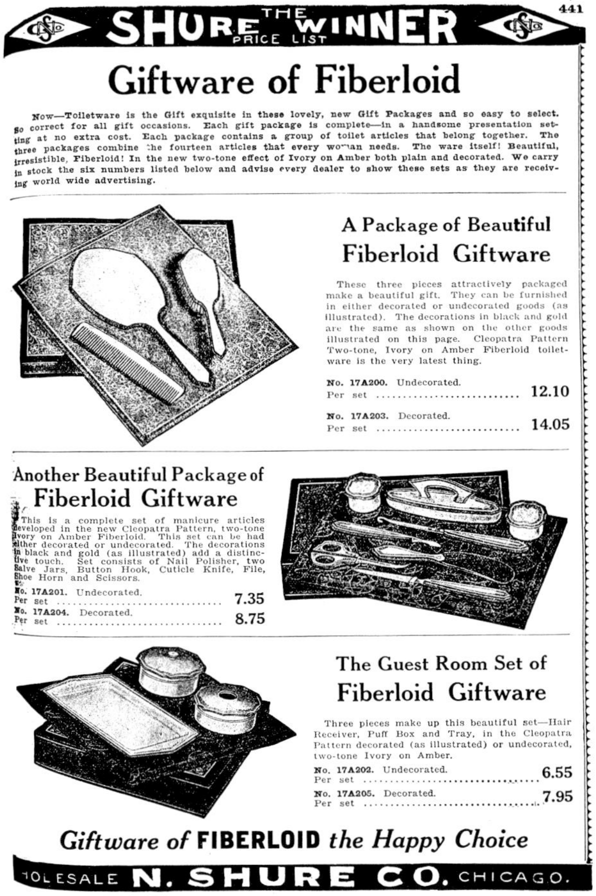

441

SHURE THE WINNER PRICE LIST

Giftware of Fiberloid

Now—Toiletware is the Gift exquisite in these lovely, new Gift Packages and so easy to select. So correct for all gift occasions. Each gift package is complete—in a handsome presentation setting at no extra cost. Each package contains a group of toilet articles that belong together. The three packages combine the fourteen articles that every woman needs. The ware itself! Beautiful, irresistible, Fiberloid! In the new two-tone effect of Ivory on Amber both plain and decorated. We carry in stock the six numbers listed below and advise every dealer to show these sets as they are receiving world wide advertising.

A Package of Beautiful Fiberloid Giftware

These three pieces attractively packaged make a beautiful gift. They can be furnished in either decorated or undecorated goods (as illustrated). The decorations in black and gold are the same as shown on the other goods illustrated on this page. Cleopatra Pattern Two-tone, Ivory on Amber Fiberloid toiletware is the very latest thing.

No. 17A200. Undecorated. Per set 12.10

No. 17A203. Decorated. Per set 14.05

Another Beautiful Package of Fiberloid Giftware

This is a complete set of manicure articles developed in the new Cleopatra Pattern, two-tone Ivory on Amber Fiberloid. This set can be had either decorated or undecorated. The decorations in black and gold (as illustrated) add a distinctive touch. Set consists of Nail Polisher, two Salve Jars, Button Hook, Cuticle Knife, File, Shoe Horn and Scissors.

No. 17A201. Undecorated. Per set 7.35

No. 17A204. Decorated. Per set 8.75

The Guest Room Set of Fiberloid Giftware

Three pieces make up this beautiful set—Hair Receiver, Puff Box and Tray, in the Cleopatra Pattern decorated (as illustrated) or undecorated, two-tone Ivory on Amber.

No. 17A202. Undecorated. Per set 6.55

No. 17A205. Decorated. Per set 7.95

Giftware of **FIBERLOID** *the Happy Choice*

WHOLESALE N. SHURE CO. CHICAGO.

1924

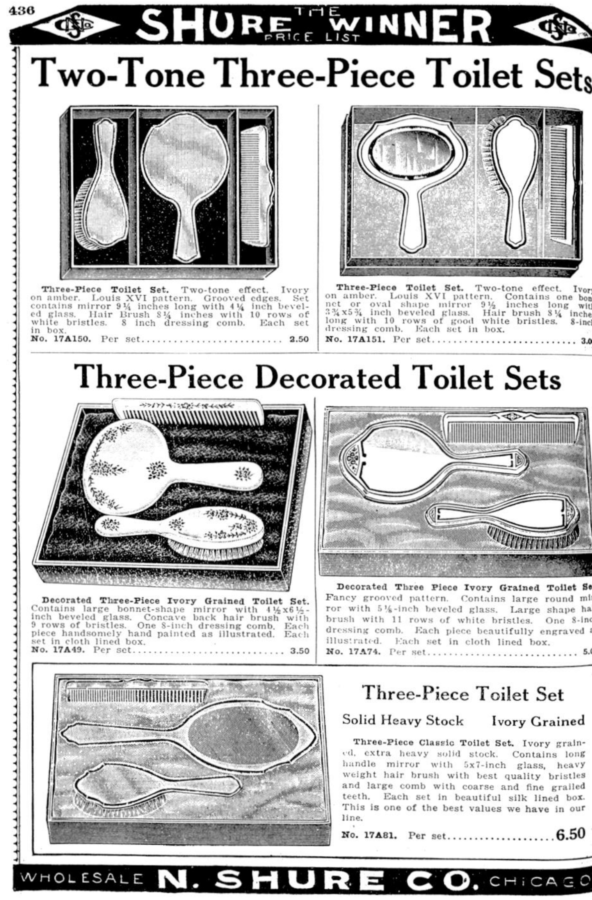

436

SHURE THE WINNER PRICE LIST

Two-Tone Three-Piece Toilet Sets

Three-Piece Toilet Set. Two-tone effect. Ivory on amber. Louis XVI pattern. Grooved edges. Set contains mirror 9¼ inches long with 4¼ inch beveled glass. Hair Brush 8¼ inches with 10 rows of white bristles. 8 inch dressing comb. Each set in box.
No. 17A150. Per set 2.50

Three-Piece Toilet Set. Two-tone effect. Ivory on amber. Louis XVI pattern. Contains one bonnet or oval shape mirror 9½ inches long with 3¾x5¾ inch beveled glass. Hair brush 8¼ inches long with 10 rows of good white bristles. 8-inch dressing comb. Each set in box.
No. 17A151. Per set 3.00

Three-Piece Decorated Toilet Sets

Decorated Three-Piece Ivory Grained Toilet Set. Contains large bonnet-shape mirror with 4½x6½-inch beveled glass. Concave back hair brush with 9 rows of bristles. One 8-inch dressing comb. Each piece handsomely hand painted as illustrated. Each set in cloth lined box.
No. 17A49. Per set 3.50

Decorated Three Piece Ivory Grained Toilet Set. Fancy grooved pattern. Contains large round mirror with 5⅛-inch beveled glass. Large shape hair brush with 11 rows of white bristles. One 8-inch dressing comb. Each piece beautifully engraved as illustrated. Each set in cloth lined box.
No. 17A74. Per set 5.00

Three-Piece Toilet Set

Solid Heavy Stock Ivory Grained

Three-Piece Classic Toilet Set. Ivory grained, extra heavy solid stock. Contains long handle mirror with 5x7-inch glass, heavy weight hair brush with best quality bristles and large comb with coarse and fine grailed teeth. Each set in beautiful silk lined box. This is one of the best values we have in our line.
No. 17A81. Per set 6.50

WHOLESALE N. SHURE CO. CHICAGO

1924

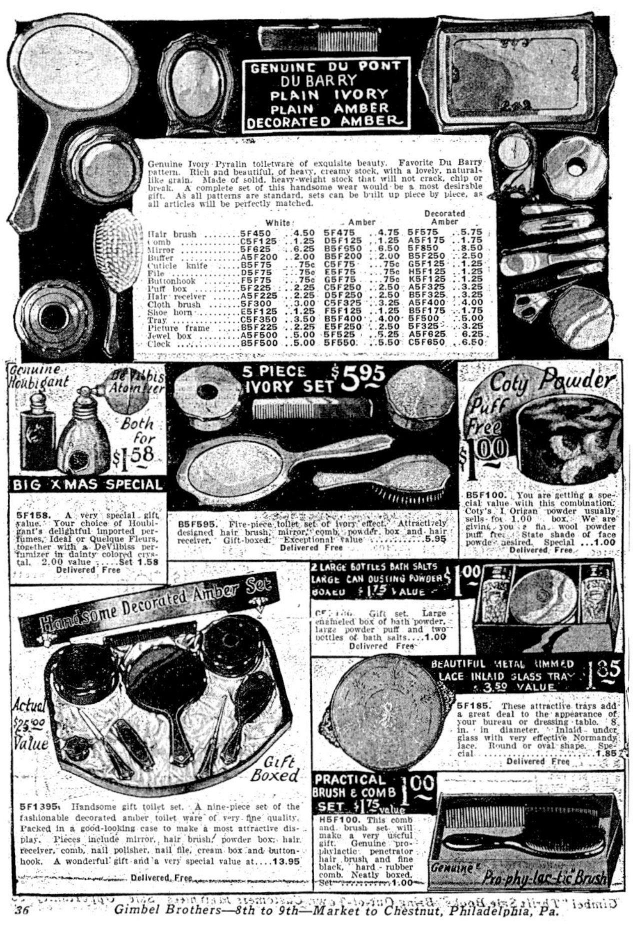

GENUINE DU PONT
DU BARRY
PLAIN IVORY
PLAIN AMBER
DECORATED AMBER

Genuine Ivory Pyralin toiletware of exquisite beauty. Favorite Du Barry pattern. Rich and beautiful, of heavy, creamy stock, with a lovely, natural-like grain. Made of solid, heavy-weight stock that will not crack, chip or break. A complete set of this handsome wear would be a most desirable gift. As all patterns are standard, sets can be built up piece by piece, as all articles will be perfectly matched.

	White		Amber		Decorated Amber	
Hair brush	5F450	4.50	5F475	4.75	5F575	5.75
Comb	C5F125	1.25	D5F125	1.25	A5F175	1.75
Mirror	5F625	6.25	B5F650	6.50	5F850	8.50
Buffer	A5F200	2.00	B5F200	2.00	B5F250	2.50
Cuticle knife	B5F75	75c	C5F75	75c	G5F125	1.25
File	D5F75	75c	E5F75	75c	H5F125	1.25
Buttonhook	F5F75	75c	G5F75	75c	K5F125	1.25
Puff box	5F225	2.25	C5F250	2.50	A5F325	3.25
Hair receiver	A5F225	2.25	D5F250	2.50	B5F325	3.25
Cloth brush	5F300	3.00	C5F325	3.25	A5F400	4.00
Shoe horn	E5F125	1.25	F5F125	1.25	B5F175	1.75
Tray	C5F350	3.50	B5F400	4.00	5F500	5.00
Picture frame	B5F225	2.25	E5F250	2.50	5F325	3.25
Jewel box	A5F500	5.00	5F525	5.25	A5F625	6.25
Clock	B5F500	5.00	5F550	5.50	C5F650	6.50

Genuine Houbigant
DeVilbis Atomizer
Both for $1.58
BIG XMAS SPECIAL

5F158. A very special gift value. Your choice of Houbigant's delightful imported perfumes, Ideal or Quelque Fleurs, together with a DeVilbiss perfumizer in dainty colored crystal. 2.00 value ......Set 1.58
Delivered Free

5 PIECE IVORY SET $5.95

B5F595. Five-piece toilet set of ivory effect. Attractively designed hair brush, mirror, comb, powder box and hair receiver. Gift-boxed. Exceptional value 5.95
Delivered Free

Coty Powder
Puff Free
$1.00

B5F100. You are getting a special value with this combination. Coty's L'Origan powder usually sells for 1.00 box. We are giving you a fine wool powder puff free. State shade of face powder desired. Special ...1.00
Delivered Free

Handsome Decorated Amber Set
Actual $25.00 Value
Gift Boxed

5F1395. Handsome gift toilet set. A nine-piece set of the fashionable decorated amber toilet ware of very fine quality. Packed in a good-looking case to make a most attractive display. Pieces include mirror, hair brush, powder box, hair receiver, comb, nail polisher, nail file, cream box and buttonhook. A wonderful gift and a very special value at....13.95
Delivered Free

2 LARGE BOTTLES BATH SALTS
LARGE CAN DUSTING POWDER
BOXED $1.75 VALUE
$1.00

Gift set. Large enameled box of bath powder, large powder puff and two bottles of bath salts....1.00
Delivered Free

BEAUTIFUL METAL RIMMED LACE INLAID GLASS TRAY $3.50 VALUE 1.85

5F185. These attractive trays add a great deal to the appearance of your bureau or dressing table. 8 in. in diameter. Inlaid under glass with very effective Normandy lace. Round or oval shape. Special 1.85
Delivered Free

PRACTICAL BRUSH & COMB SET $1.75 value 1.00

H5F100. This comb and brush set will make a very useful gift. Genuine prophylactic penetrator hair brush and fine black, hard-rubber comb. Neatly boxed. Set 1.00

Genuine "Pro-phy-lac-tic" Brush

36 Gimbel Brothers—8th to 9th—Market to Chestnut, Philadelphia, Pa.

1924

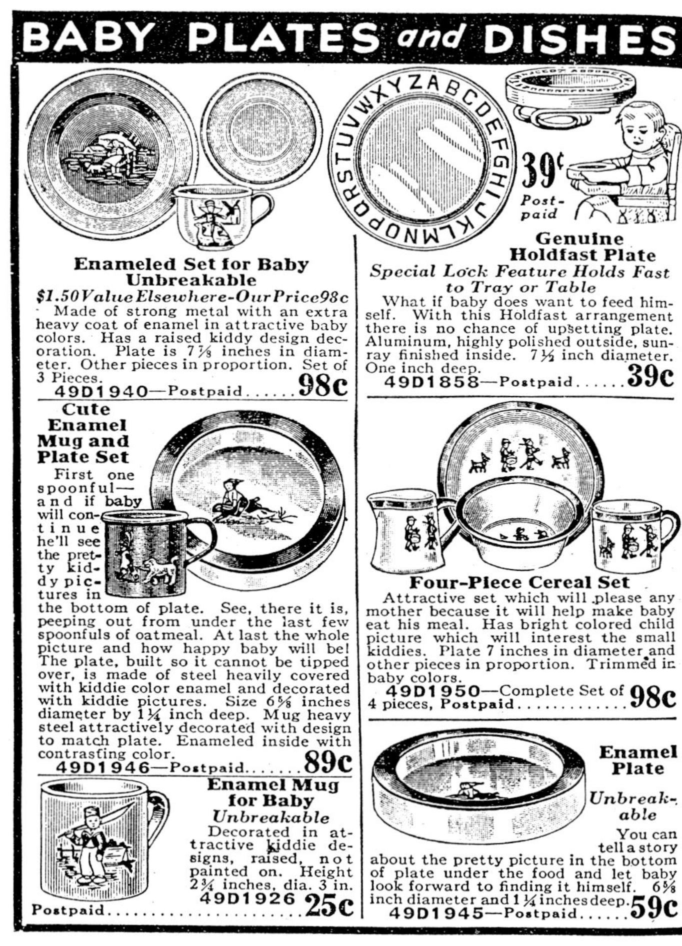

BABY PLATES and DISHES

ABCDEFGHIJKLMNOPQRSTUVWXYZ

Enameled Set for Baby Unbreakable

$1.50 Value Elsewhere-Our Price 98c

Made of strong metal with an extra heavy coat of enamel in attractive baby colors. Has a raised kiddy design decoration. Plate is 7⅛ inches in diameter. Other pieces in proportion. Set of 3 Pieces.
49D1940—Postpaid 98c

39¢ Postpaid

Genuine Holdfast Plate

Special Lock Feature Holds Fast to Tray or Table

What if baby does want to feed himself. With this Holdfast arrangement there is no chance of upsetting plate. Aluminum, highly polished outside, sunray finished inside. 7½ inch diameter. One inch deep.
49D1858—Postpaid 39c

Cute Enamel Mug and Plate Set

First one spoonful—and if baby will continue he'll see the pretty kiddy pictures in the bottom of plate. See, there it is, peeping out from under the last few spoonfuls of oatmeal. At last the whole picture and how happy baby will be! The plate, built so it cannot be tipped over, is made of steel heavily covered with kiddie color enamel and decorated with kiddie pictures. Size 6⅝ inches diameter by 1¼ inch deep. Mug heavy steel attractively decorated with design to match plate. Enameled inside with contrasting color.
49D1946—Postpaid 89c

Four-Piece Cereal Set

Attractive set which will please any mother because it will help make baby eat his meal. Has bright colored child picture which will interest the small kiddies. Plate 7 inches in diameter and other pieces in proportion. Trimmed in baby colors.
49D1950—Complete Set of 4 pieces, Postpaid 98c

Enamel Mug for Baby

Unbreakable

Decorated in attractive kiddie designs, raised, not painted on. Height 2¾ inches, dia. 3 in.
49D1926
Postpaid 25c

Enamel Plate

Unbreakable

You can tell a story about the pretty picture in the bottom of plate under the food and let baby look forward to finding it himself. 6⅝ inch diameter and 1¼ inches deep.
49D1945—Postpaid 59c

1930

114 CELLULOID AND BISQUE DOLLS

Feather Dressed Celluloid Doll

Flesh colored celluloid doll, painted wavy hair and large painted eyes, gold glass beads around neck and hanging from ears. Feather dress in assorted colors, metallic hat and celluloid cane. Height of doll, 7 inches.

No. 08/29. Per Dozen.........**$1.20**
Per Gross **13.50**

DOLL IN CARRIAGE

Wicker carriage with wooden wheels. Rayon drape hanging from shade. Size 5 inches long and 5 inches high. Contains a dressed bisque doll with movable arms and legs.

No. 11/68. Per Dozen.........**$1.20**
Per Gross **13.50**

Feather Doll

Celluloid Doll, 6 ins. high, movable hands—with ostrich feather dress in assorted colors—height with dress, 9 inches.

No. 028/0. Per Dozen.........**$1.10**
Per Gross **12.00**

CELLULOID DOLL

Flesh colored full body, painted wavy hair, big black eyes, painted eyelashes and moving arms; 6 inches long.

No. 19/40.
Per Dozen **$0.50**
Per Gross **5.50**

JUMPING RABBIT

Made of fur. When the rubber bulb is pressed the rabbit jumps and squeaks. Size 4½ inches long.

No. 6/22. Per Dozen........**$ 0.90**
Per Gross **10.00**

BISQUE DOLL

Painted hair and features, jointed arms and rayon dress in pink or blue. Six inches high and packed 1 dozen in box, assorted. Very life-like in appearance.

No. 14/54.
Per Dozen **$0.50**
Per Gross **5.50**

SLACK MFG. CO. — 124-126 W. LAKE ST. CHICAGO, ILL.

1932

12 LAPEL NOVELTIES

ASSORTED CELLULOID FLOWERS

An assortment of different flowers in assorted colors. Pin on back. Each on card. Packed 1 gross in box (NO LESS SOLD).

No. 36/06. Per Gross.........**$1.25**

SCOTTY DOG PIN

Made of composition, actual size of illustration, in white with black or brown spots. Pin on back.

No. 49/21. Per Gross..........**$1.25**

CELLULOID ROSE PIN

Two layer red rose, leaf back, glass bead center. Packed 1 dozen on card, 1 gross to box. No less than gross sold.

No. 4/27. Per Gross...........**$0.75**

BUTTERFLY PIN

Made of metal in assorted true colors. Easily attached to men's coat lapel or ladies' waist. An attractive pin for favors or as an identification marker. Sold in gross lots only.

No. 43/6. Per Gross...........**$1.10**

CELLULOID ROSE PINS

Miniature Red Rose, streamers are blue and white, both made of celluloid. Length 1¼ inches. One dozen on card, gross in package, no less sold.

No. 202/8. Per Gross..**$1.00**

FOOTBALL CHARM

Made of moulded celluloid in brown finish. Size ¾ inch long. Has ring with assorted colored cord to fasten to lapel. Packed 1 gross to box (NO LESS SOLD).

No. 4/29. Per Gross...........**$1.25**

CELLULOID ROSETTE PIN

Three layer red flower with saw tooth edge, leaf back, glass bead center. Size of flower ¾ inch in diameter. Each on card. Packed gross in box—no less sold.

No. 26/56. Per Gross.........**$1.25**

METAL FOOTBALL CHARM

Made of metal, gilt finish, size 1$\frac{1}{16}$ inches long. Has ring on top to fasten to ribbon badge or cane. Packed 100 to box (NO LESS SOLD).

No. 29/51. Per Gross........**$ 1.75**
Per 1,000 .... **11.00**

CELLULOID DOLL PIN

Little celluloid doll with a celluloid leaf in back in assorted colors. Pin on back. Height, 1½ in. Packed 1 gross to box (no less sold).

No. 18/1. Per Gross..........**$1.10**

LION CHARM

Made of moulded celluloid in ivory finish. Size 1⅛ inches long. Has ring with assorted colored cord to fasten to lapel. Packed 1 gross to box (NO LESS SOLD).

No. 12/46. Per Gross.........**$1.25**

SLACK MFG. CO. — 124-126 W. LAKE STREET CHICAGO, ILL.

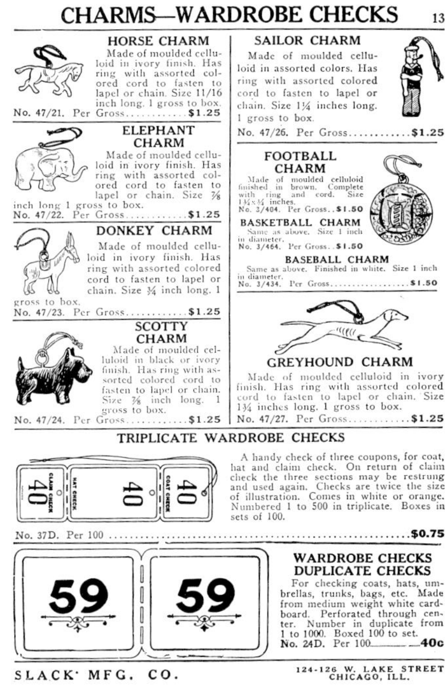

CHARMS—WARDROBE CHECKS 13

HORSE CHARM

Made of moulded celluloid in ivory finish. Has ring with assorted colored cord to fasten to lapel or chain. Size 11/16 inch long. 1 gross to box.

No. 47/21. Per Gross............**$1.25**

SAILOR CHARM

Made of moulded celluloid in assorted colors. Has ring with assorted colored cord to fasten to lapel or chain. Size 1¼ inches long. 1 gross to box.

No. 47/26. Per Gross............**$1.25**

ELEPHANT CHARM

Made of moulded celluloid in ivory finish. Has ring with assorted colored cord to fasten to lapel or chain. Size ⅞ inch long 1 gross to box.

No. 47/22. Per Gross............**$1.25**

FOOTBALL CHARM

Made of moulded celluloid finished in brown. Complete with ring and cord. Size 1¾x¾ inches.

No. 3/404. Per Gross..**$1.50**

BASKETBALL CHARM

Same as above. Size 1 inch in diameter.

No. 3/464. Per Gross..**$1.50**

BASEBALL CHARM

Same as above. Finished in white. Size 1 inch in diameter.

No. 3/434. Per Gross...................**$1.50**

DONKEY CHARM

Made of moulded celluloid in ivory finish. Has ring with assorted colored cord to fasten to lapel or chain. Size ¾ inch long. 1 gross to box.

No. 47/23. Per Gross............**$1.25**

SCOTTY CHARM

Made of moulded celluloid in black or ivory finish. Has ring with assorted colored cord to fasten to lapel or chain. Size ⅞ inch long. 1 gross to box.

No. 47/24. Per Gross............**$1.25**

GREYHOUND CHARM

Made of moulded celluloid in ivory finish. Has ring with assorted colored cord to fasten to lapel or chain. Size 1¾ inches long. 1 gross to box.

No. 47/27. Per Gross............**$1.25**

TRIPLICATE WARDROBE CHECKS

A handy check of three coupons, for coat, hat and claim check. On return of claim check the three sections may be restrung and used again. Checks are twice the size of illustration. Comes in white or orange. Numbered 1 to 500 in triplicate. Boxes in sets of 100.

No. 37D. Per 100 ....**$0.75**

WARDROBE CHECKS DUPLICATE CHECKS

For checking coats, hats, umbrellas, trunks, bags, etc. Made from medium weight white cardboard. Perforated through center. Number in duplicate from 1 to 1000. Boxed 100 to set.

No. 24D. Per 100.............**40c**

SLACK MFG. CO. — 124-126 W. LAKE STREET CHICAGO, ILL.

1936

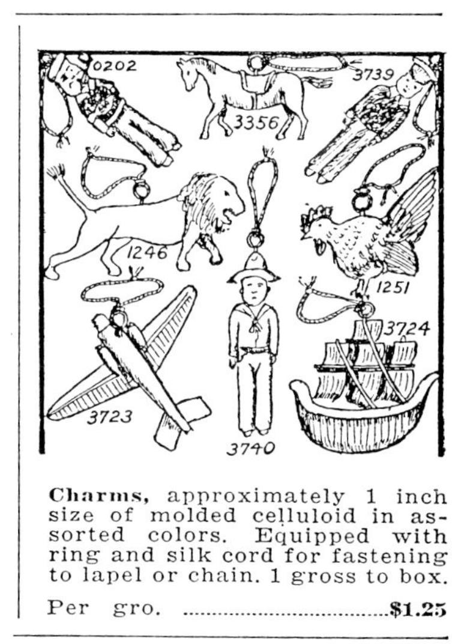

Charms, approximately 1 inch size of molded celluloid in assorted colors. Equipped with ring and silk cord for fastening to lapel or chain. 1 gross to box.

Per gro. **$1.25**

1932

114 DOLLS—TOYS

Feather Dressed Celluloid Doll

Flesh colored celluloid doll, painted wavy hair and large painted eyes, gold glass beads around neck and hanging from ears. Feather dress in assorted colors, metallic hat and celluloid cane. Height of doll, 7 inches.

No. 08/29. Per Dozen.........**$1.20**
Per Gross **13.50**

FLYING BIRD
With Hummer

Three color flying bird with a metal contraption which produces the sound of the humming bird. Length of bird and tail 10½ inches. Comes with a long bright colored spiral wound stick. When swung in air the tail revolves giving the appearance of flying. Come packed 3 dozen in box. (NO LESS SOLD.)

No. 517B. Per Gross....**$3.00**
Per Box of 3 Dozen...... **.80**

BOW AND ARROW

Consists of a 24-inch bow, a harmless rubber tip suction arrow and two paper targets. It is fastened to a decorated cardboard and comes 1 dozen to package.

No. 9/15B.
Per Dozen..**$0.50**
Per Gross... **5.50**

CELLULOID DOLL

Flesh colored full body, painted wavy hair, big black eyes, painted eyelashes and moving arms; 6 inches long.

No. 19/40.
Per Dozen **$0.50**
Per Gross **5.50**

PENCIL SHARPENER

Made of metal, a reproduction of an alarm clock, with a pencil sharpener at the bottom. Size 1½ inches in diameter. Very Novel. Packed one dozen in box.

No. 16/66. Per Dozen...**$0.50**
Per Gross **5.50**

PUZZLES

Made of heavy colored paper with glass top. Comes assorted in 6 different styles. Object of puzzles is to get the balls in the eyes of the figures. Size 2¼ inches in diameter. Packed 1 gross in box. (NO LESS SOLD.)

No. 21/02P. Per Gross..............**$1.10**

SLACK MFG. CO. — 124-126 W. LAKE ST. CHICAGO, ILL.

MECHANICAL TOYS AND WATCHES 115

TOY BRACELET WATCH

Plain metal case, printed dial, celluloid crystal, moving hands and leather strap. A very attractive toy. Each in a box.

No. 29/3. Per Dozen.............**$0.60**
Per Gross **6.75**

MECHANICAL HULA DANCER

Brown painted celluloid body, head and arms. Legs of metal. Dressed in rayon fringe skirt. Has powerful spring and when wound up it does the Hula dance. Size 6¾ inches high. Individually boxed.

No. 21/67. Per Dozen....**$ 1.25**
Per Gross **14.00**

MECHANICAL TOY

Celluloid Jockey and Indian on horse, on metal wheels. Good strong spring motor. Size 5 inches overall.

No. 16T79. Per Dozen......**$0.80**
Per Gross **9.50**

MECHANICAL CLOWN

Made of metal and painted in brilliant colors. Has a powerful spring and when wound it runs around in a circle and the cube on top keeps revolving. Size 6½ inches high overall. Individually boxed.

No. 9/75. Per Dozen.....**$ 1.00**
Per Gross **11.50**

MECHANICAL DANCING COUPLE

Made of celluloid, nicely decorated. When wound up they dance around. Good spring motor. 4½ inches high. Each in box.

No. 49/31. Per Dozen ....**$ 1.25**
Per Gross **14.00**

JOKER CAR

A mechanical toy made of metal, painted in vivid colors. Has a very powerful spring and when wound it runs backward, forward, sidewise and in a circle. A very comical item that will cause much laughter. Size 3 inches long and 3½ inches high overall. Individually boxed.

No. 2/89. Per Dozen.....**$ 1.10**
Per Gross **12.00**

MECHANICAL CRAWLING BABY

Head, arms and legs made of celluloid. Dressed in a printed cloth romper. Crawls when wound up. Good spring motor. Measures 3 x 4 inches.

No. 28/74. Per Dozen ....**$ 1.25**
Per Gross **14.00**

1936

330

GELLMAN BROS. MINNEAPOLIS, MINN.

NOVELTY FEATHER DRESSED DOLLY and FUR TRIMMED MONKEY CANES

No. 3714—Rainbow Colored Jumping Monkey Assortment. Includes an assortment of monkeys made with composition heads in assorted colors with painted features and assorted colored metallic high hats. Height, 6½ inches. Complete with Jap crook canes.
Per 100 $3.50

No. 3712—Dolly Cane. Consists of 5-inch celluloid vamp doll with beautiful feather dress in striking colors. Gilt painted wavy hair effect and fancy Indian style feather head-dress. Complete with Jap crook canes.
Per 100 $4.35

No. 3719—South Sea Island Baby Dolly Cane. This is one of our best sellers. Includes 7-inch celluloid vamp doll painted to represent a South Sea Island dancing girl dressed with brightly colored feathers and beads and spangles. Complete with Jap crook canes.
Per 100 $6.25

No. 3711—Dolly Cane. Doll is made of celluloid and strikingly dressed in beautifully colored Butterfly effect feather dress and has gilt painted wavy hair with lace effect waist band. Height of doll, 5 in. Complete with Jap crook canes.
Per 100 $3.85

No. 3713—Rainbow Colored Jumping Monkey Assortment. Monkeys are made with composition heads in assorted colors having painted features and brightly colored metallic high hats. Height, 8 inches. Complete with Jap crook canes.
Per 100 $5.20

No. 3718—Dolly Cane. Beautiful 7-inch celluloid vamp doll dressed in Butterfly effect assorted bright colored feathers with gilt painted wavy hair and gilt bead ear-rings and necklace. Complete with Jap crook canes.
Per 100.......... $5.90

No. 3717 — Dolly Cane. Vamp style 7-inch celluloid doll dressed in beautifully colored feathers with gilt bead ear-rings and necklace. Attractive gilt painted wavy hair effect with fancy Indian style feather head-dress. Complete with Jap crook canes.
Per 100.......... $6.25

No. 3707—Dolly Cane. Consists of a 7-inch celluloid vamp doll handsomely dressed in attractively colored Butterfly effect feathers with gilt painted wavy hair and lace effect waist band. Complete with Jap crook canes.
Per 100.......... $5.20

No. 3722—Dolly Cane. Beautiful 7-inch celluloid vamp doll handsomely dressed in colored feathers with gilt bead ear-rings and necklace and gilt painted wavy hair effect. Complete with Jap crook canes.
Per 100.......... $6.40

No. 3724—Dolly Cane Newest vamp style 7-inch celluloid doll dressed in colored feathers with gilt bead ear-rings and necklace. Features colored cardboard high-hat and celluloid cane with gilt painted wavy hair effect. Complete with Jap crook canes.
Per 100............... $7.00

No. 3723—Dolly Cane. Handsome 8-inch celluloid vamp doll dressed in beautifully colored feathers with gilt bead ear-rings and necklace. Has gilt painted wavy hair effect. Complete with Jap crook canes.
Per 100............... $7.25

WHOLESALE PREMIUM MERCHANDISE & NOVELTIES

1936

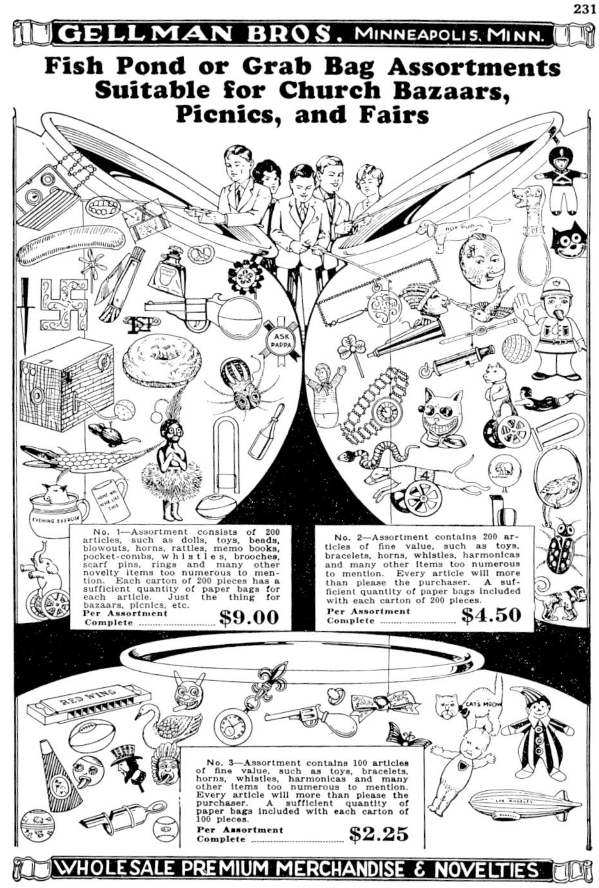

231

GELLMAN BROS. MINNEAPOLIS, MINN.

Fish Pond or Grab Bag Assortments Suitable for Church Bazaars, Picnics, and Fairs

No. 1—Assortment consists of 200 articles, such as dolls, toys, beads, blowouts, horns, rattles, memo books, pocket-combs, whistles, brooches, scarf pins, rings and many other novelty items too numerous to mention. Each carton of 200 pieces has a sufficient quantity of paper bags for each article. Just the thing for bazaars, picnics, etc.
Per Assortment Complete $9.00

No. 2—Assortment contains 200 articles of fine value, such as toys, bracelets, horns, whistles, harmonicas and many other items too numerous to mention. Every article will more than please the purchaser. A sufficient quantity of paper bags included with each carton of 200 pieces.
Per Assortment Complete $4.50

No. 3—Assortment contains 100 articles of fine value, such as toys, bracelets, horns, whistles, harmonicas and many other items too numerous to mention. Every article will more than please the purchaser. A sufficient quantity of paper bags included with each carton of 100 pieces.
Per Assortment Complete $2.25

WHOLESALE PREMIUM MERCHANDISE & NOVELTIES

1936

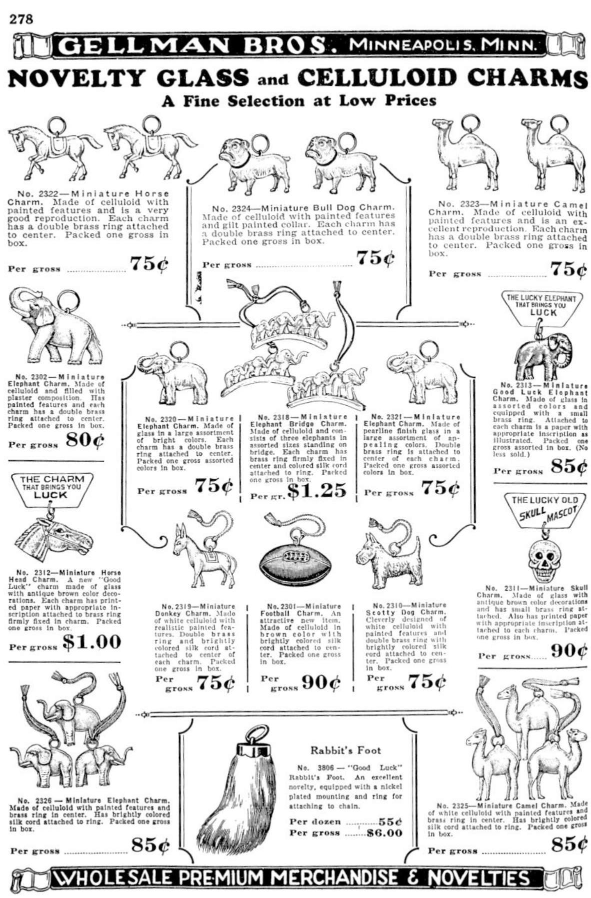

278

GELLMAN BROS. MINNEAPOLIS, MINN.

NOVELTY GLASS and CELLULOID CHARMS

A Fine Selection at Low Prices

No. 2322—Miniature Horse Charm. Made of celluloid with painted features and is a very good reproduction. Each charm has a double brass ring attached to center. Packed one gross in box.
Per gross .... 75¢

No. 2324—Miniature Bull Dog Charm. Made of celluloid with painted features and gilt painted collar. Each charm has a double brass ring attached to center. Packed one gross in box.
Per gross .... 75¢

No. 2323—Miniature Camel Charm. Made of celluloid with painted features and is an excellent reproduction. Each charm has a double brass ring attached to center. Packed one gross in box.
Per gross .... 75¢

No. 2302—Miniature Elephant Charm. Made of celluloid and filled with plaster composition. Has painted features and each charm has a double brass ring attached to center. Packed one gross in box.
Per gross 80¢

THE LUCKY ELEPHANT THAT BRINGS YOU LUCK

No. 2313—Miniature Good Luck Elephant Charm. Made of glass in assorted colors and equipped with a small brass ring. Attached to each charm is a paper with appropriate inscription as illustrated. Packed one gross assorted in box. (No less sold.)
Per gross 85¢

No. 2320—Miniature Elephant Charm. Made of glass in a large assortment of bright colors. Each charm has a double brass ring attached to center. Packed one gross assorted colors in box.
Per gross 75¢

No. 2318—Miniature Elephant Bridge Charm. Made of celluloid and consists of three elephants in assorted sizes standing on bridge. Each charm has brass ring firmly fixed in center and colored silk cord attached to ring. Packed one gross in box.
Per gr. $1.25

No. 2321—Miniature Elephant Charm. Made of pearline finish glass in a large assortment of appealing colors. Double brass ring is attached to center of each charm. Packed one gross assorted colors in box.
Per gross 75¢

THE CHARM THAT BRINGS YOU LUCK

No. 2312—Miniature Horse Head Charm. A new "Good Luck" charm made of glass with antique brown color decorations. Each charm has printed paper with appropriate inscription attached to brass ring firmly fixed in charm. Packed one gross in box.
Per gross $1.00

THE LUCKY OLD SKULL MASCOT

No. 2311—Miniature Skull Charm. Made of glass with antique brown color decorations and has small brass ring attached. Also has printed paper with appropriate inscription attached to each charm. Packed one gross in box.
Per gross...... 90¢

No. 2319—Miniature Donkey Charm. Made of white celluloid with realistic painted features. Double brass ring and brightly colored silk cord attached to center of each charm. Packed one gross in box.
Per gross 75¢

No. 2301—Miniature Football Charm. An attractive new item. Made of celluloid in brown color with brightly colored silk cord attached to center. Packed one gross in box.
Per gross 90¢

No. 2310—Miniature Scotty Dog Charm. Cleverly designed of white celluloid with painted features and double brass ring with brightly colored silk cord attached to center. Packed one gross in box.
Per gross 75¢

No. 2326 — Miniature Elephant Charm. Made of celluloid with painted features and brass ring in center. Has brightly colored silk cord attached to ring. Packed one gross in box.
Per gross .... 85¢

Rabbit's Foot

No. 3806 — "Good Luck" Rabbit's Foot. An excellent novelty, equipped with a nickel plated mounting and ring for attaching to chain.
Per dozen ....55¢
Per gross $6.00

No. 2325—Miniature Camel Charm. Made of white celluloid with painted features and brass ring in center. Has brightly colored silk cord attached to ring. Packed one gross in box.
Per gross .... 85¢

WHOLESALE PREMIUM MERCHANDISE & NOVELTIES

1936

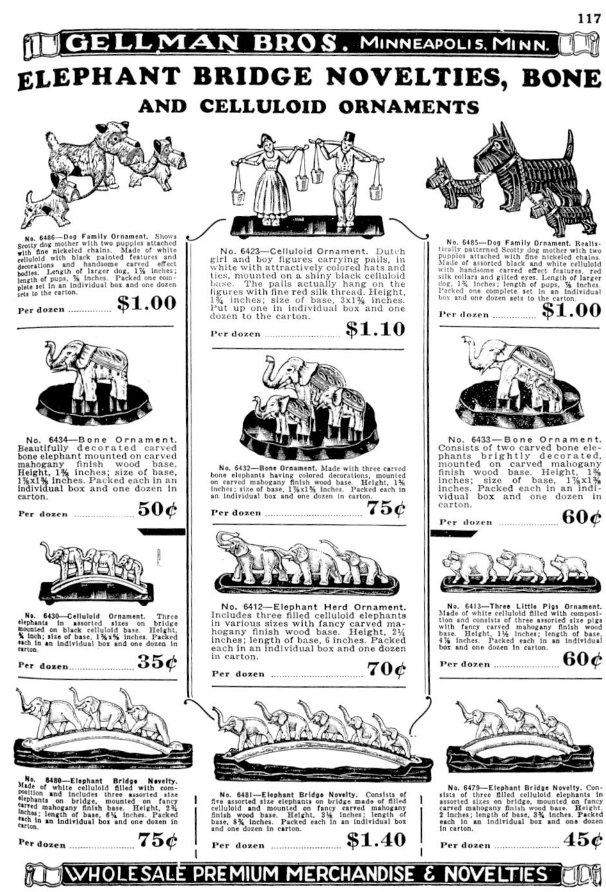

117

GELLMAN BROS. MINNEAPOLIS, MINN.

ELEPHANT BRIDGE NOVELTIES, BONE AND CELLULOID ORNAMENTS

No. 6486—Dog Family Ornament. Shows Scotty dog mother with two puppies attached with fine nickeled chains. Made of white celluloid with black painted features and decorations and handsome carved effect bodies. Length of larger dog, 1⅞ inches; length of pups, ⅞ inches. Packed one complete set in an individual box and one dozen sets to the carton.
Per dozen $1.00

No. 6423—Celluloid Ornament. Dutch girl and boy figures carrying pails, in white with attractively colored hats and ties, mounted on a shiny black celluloid base. The pails actually hang on the figures with fine red silk thread. Height, 1¾ inches; size of base, 3x1⅜ inches. Put up one in individual box and one dozen to the carton.
Per dozen $1.10

No. 6485—Dog Family Ornament. Realistically patterned Scotty dog mother with two puppies attached with fine nickeled chains. Made of assorted black and white celluloid with handsome carved effect features, red silk collars and gilted eyes. Length of larger dog, 1¾ inches; length of pups, ⅞ inches. Packed one complete set in an individual box and one dozen sets to the carton.
Per dozen $1.00

No. 6434—Bone Ornament. Beautifully decorated carved bone elephant mounted on carved mahogany finish wood base. Height, 1⅜ inches; size of base, 1⅞x1⅜ inches. Packed each in an individual box and one dozen in carton.
Per dozen .... 50¢

No. 6432—Bone Ornament. Made with three carved bone elephants having colored decorations, mounted on carved mahogany finish wood base. Height, 1⅜ inches; size of base, 1⅞x1⅜ inches. Packed each in an individual box and one dozen in carton.
Per dozen .... 75¢

No. 6433—Bone Ornament. Consists of two carved bone elephants brightly decorated, mounted on carved mahogany finish wood base. Height, 1⅜ inches; size of base, 1⅞x1⅜ inches. Packed each in an individual box and one dozen in carton.
Per dozen .... 60¢

No. 6430—Celluloid Ornament. Three elephants in assorted sizes on bridge mounted on black celluloid base. Height, ¾ inch; size of base, 1⅜x⅝ inches. Packed each in an individual box and one dozen in carton.
Per dozen.................. 35¢

No. 6412—Elephant Herd Ornament. Includes three filled celluloid elephants in various sizes with fancy carved mahogany finish wood base. Height, 2¼ inches; length of base, 6 inches. Packed each in an individual box and one dozen in carton.
Per dozen .... 70¢

No. 6413—Three Little Pigs Ornament. Made of white celluloid filled with composition and consists of three assorted size pigs with fancy carved mahogany finish wood base. Height, 1½ inches; length of base, 4⅞ inches. Packed each in an individual box and one dozen in carton.
Per dozen .... 60¢

No. 6480—Elephant Bridge Novelty. Made of white celluloid filled with composition and includes three assorted size elephants on bridge, mounted on fancy carved mahogany finish base. Height, 2¾ inches; length of base, 6¼ inches. Packed each in an individual box and one dozen in carton.
Per dozen .... 75¢

No. 6481—Elephant Bridge Novelty. Consists of five assorted size elephants on bridge made of filled celluloid and mounted on fancy carved mahogany finish wood base. Height, 3⅛ inches; length of base, 8¾ inches. Packed each in an individual box and one dozen in carton.
Per dozen $1.40

No. 6479—Elephant Bridge Novelty. Consists of three filled celluloid elephants in assorted sizes on bridge, mounted on fancy carved mahogany finish wood base. Height, 2 inches; length of base, 3¾ inches. Packed each in an individual box and one dozen in carton.
Per dozen .... 45¢

WHOLESALE PREMIUM MERCHANDISE & NOVELTIES

1936

308

GELLMAN BROS. MINNEAPOLIS, MINN.

COMIC CELLULOID BUTTONS AND NOVELTY BADGES

Comic Celluloid Buttons

No. 4005—Celluloid Buttons with Comic Sayings. Size 50 ligne ($1\frac{1}{4}$ inches in diameter). Each package contains a large variety of buttons with snappy sayings. Lettering in black, with fancy design red border and stick pin in back, 100 assorted buttons to package. No less sold.

Per 100 77c

Per 1000 $7.50

Per 100.................77¢ Per 1,000.............$7.50

COMIC RIBBON BADGE ASSORTMENT

Comic Badge Assortment

No. 5482—This assortment consists of 12 of the latest novelties made up with 50 ligne Celluloid Buttons with comic sayings, attached to silk faced ribbon. One dozen assorted badges to package.

Per Dozen 35c

Per Gross $3.85

Per dozen..........................35¢ Per gross.......................$3.85

1936

115

GELLMAN BROS. MINNEAPOLIS, MINN.

"What-not" and Curio Cabinet Novelties

Made of Glass and Celluloid

No. 6419 — Celluloid Ornament. Consists of rooster on tree and two chickens all beautifully designed, mounted on black celluloid base. Height, $2\frac{1}{2}$ inches; size of base, $2\text{x}1\frac{3}{8}$ inches. Packed each in an individual box and one dozen in carton.

Per dozen.................75¢

No. 6478—Glass Ornament Assortment. Includes a large variety of miniature glass animals in assorted colors. Each animal is supplied with a carved mahogany finish wood base. Packed one dozen in a partitioned box.

Per dozen..........30¢
Per gross..........$3.40

No. 6417 — Celluloid Ornament. Large stork with spread wings painted in bright colors with gilt tortoise mounted on circular black celluloid base. Height, $2\frac{3}{8}$ inches; diameter of base, $1\frac{1}{2}$ inches. Packed each in an individual box and one dozen in carton.

Per dozen.........$1.10

No. 6426 — Celluloid Ornament. Beautifully designed reindeer and doe with painted features mounted on black celluloid base. Height, $1\frac{1}{2}$ inches; size of base, $1\frac{1}{2}\text{x}1\frac{1}{8}$ inches. Packed each in an individual box and one dozen in carton.

Per dozen.................60¢

No. 6414—Glass Ornament Assortment. Consists of miniature glass animals made in very bright colors and beautifully designed. Each animal is supplied with a carved mahogany finish wood base. Put up each in an individual box complete with base. Packed one dozen assorted in a carton.

Per dozen..........65¢

No. 6431 — Celluloid Ornament. Two acrobatic elephants with painted features and with one seated on the other, mounted on black celluloid base. Height, $1\frac{1}{8}$ in.; size of base, $1\frac{1}{8}\text{x}\frac{5}{8}$ inches. Packed each in an individual box and one dozen in carton.

Per dozen.................35¢

No. 6415—Glass Ornament Assortment. This is one of our best assortments and includes a large variety of brightly colored miniature glass animals. Each supplied with a carved mahogany finish wood base. Put up each in an individual box complete with base. Packed one dozen assorted in a carton.

Per dozen..........65¢

No. 6487 — Celluloid Ornament. A prancing pony with cowboy rider in white, mounted on shiny black celluloid base. Neckerchief of rider is painted bright red. Height, $2\frac{1}{2}$ inches; size of base, $2\frac{1}{4}\text{x}1\frac{1}{4}$ inches. Put up each in an individual box and one dozen in carton.

Per dozen $1.00

No. 6442—Celluloid Ornament. Beautifully designed chicken and rooster with painted features mounted on black celluloid base. Height, $1\frac{1}{8}$ inches; size of base, $1\frac{5}{8}\text{x}1\frac{1}{8}$ inches. Packed each in an individual box and one dozen in carton.

Per dozen ....45¢

No. 6436—Celluloid Ornament. Two beautifully designed Scotty dogs in jet black mounted on white celluloid base. Height, $1\frac{5}{8}$ inches; size of base, $2\frac{5}{8}\text{x}1\frac{1}{2}$ inches. Packed each in an individual box and one dozen in carton.

Per dozen ....85¢

No. 6421 — Celluloid Ornament. An appealing design showing horse and rider jumping fence, in white with bright red hunter's coat and black hat. Mounted on black celluloid base. Height, $1\frac{7}{8}$ inches; size of base $2\frac{3}{8}\text{x}1\frac{1}{4}$ inches. Put up each in an individual box and one dozen in carton.

Per doz. $1.00

WHOLESALE PREMIUM MERCHANDISE & NOVELTIES

1936

116

GELLMAN BROS. MINNEAPOLIS, MINN.

Novelty Celluloid Figures Suitable for "What-not" Shelves

No. 6437—Celluloid Ornament. Consists of two dogs with painted features seated under bright colored parasol, mounted on black celluloid base. Height, $2\frac{3}{8}$ inches; size of base, $1\frac{5}{8}\text{x}\frac{7}{8}$ inches. Packed each in an individual box and one dozen in carton.

Per dozen....................60¢

No. 6418 — Celluloid Ornament. A rooster and two chickens, perfectly designed, mounted on black celluloid base. Height, $1\frac{3}{8}$ inches; size of base, $2\frac{7}{8}\text{x}1\frac{5}{8}$ inches. Packed each in an individual box and one dozen in carton.

Per dozen..........................90¢

No. 6439—Celluloid Ornament. Consists of comical mouse painted in bright colors with painted features, mounted on black celluloid base. Height, 2 inches; size of base, $1\frac{5}{8}\text{x}1\frac{1}{8}$ inches. Packed each in an individual box and one dozen in carton.

Per dozen....................55¢

No. 6440—Celluloid Ornament. Rooster standing on roosting post and chicken all mounted on black celluloid base. Height, $1\frac{3}{4}$ inches; size of base, $2\text{x}1\frac{1}{4}$ inches. Packed each in an individual box and one dozen in carton.

Per dozen....................60¢

No. 6429—Celluloid Ornament. Consists of celluloid cat and kitten seated on black celluloid base. Height, $1\frac{1}{8}$ inches; size of base, $1\frac{5}{8}\text{x}\frac{7}{8}$ inches. Packed each in an individual box and one dozen in carton.

Per dozen..........................50¢

No. 6446—Celluloid Ornament. Consists of chicken standing on brightly colored celluloid chair. Height, $1\frac{7}{8}$ inches; width of chair, $1\frac{1}{8}$ inches. Packed each in an individual box and one dozen in carton.

Per dozen....................45¢

No. 6420—Celluloid Ornament. Two celluloid fighting roosters perfectly designed and mounted on black celluloid base. Height, $1\frac{1}{4}$ inches; size of base, $3\text{x}1\frac{1}{4}$ inches. Packed each in an individual box and one dozen in carton.

Per dozen..........................90¢

No. 6435—Celluloid Ornament. Beautifully designed butterfly girl dancer mounted on black celluloid base. Height, $2\frac{3}{8}$ inches; size of base, $1\frac{3}{4}\text{x}1\frac{3}{4}$ inches. Packed each in an individual box and one dozen in carton.

Per dozen..........60¢

No. 6424 — Celluloid Ornament. Attractively designed dog wearing high hat and with cane in mouth, mounted on black celluloid base. Height, $1\frac{3}{8}$ inches; size of base, $1\frac{3}{8}\text{x}\frac{7}{8}$ inches. Packed each in an individual box and one dozen in carton.

Per dozen.........60¢

No. 6428 — Celluloid Ornament. Consists of two celluloid elephants in different sizes mounted on black celluloid base. Height, 1 inch; size of base, $1\frac{1}{2}\text{x}1\frac{1}{8}$ inches. Packed each in an individual box and one dozen in carton.

Per dozen.........60¢

No. 6416—Glass Ornament. Consists of tropical tree made of glass mounted on fancy carved black color wood base with cream colored glass elephant. Height, $3\frac{3}{4}$ inches; size of base, $2\frac{3}{4}\text{x}1\frac{1}{2}$ inches. Packed each in an individual box and one dozen in carton.

Per dozen.... $1.00

WHOLESALE PREMIUM MERCHANDISE & NOVELTIES

1936

PHOTO STANDS.

7668 7669

7668 Fancy shape **basket style photo holder**, made with large brocaded cloth covered wood base with embossed flexible **celluloid** sides, hand painted flower design, gilt edges. Size 7x3 inches. Doz...........3 50

7669 Celluloid **photo holder**, embossed gilt metal stand feet, wooden base, **moire silk covered** neatly tucked and gathered, hand painted sides, pinked edges, gilt trimmed. Size $3\frac{1}{4}\text{x}7\text{x}8$ inches...........Doz. 4 00

7670 **Photo holder**, tasteful novelty, made of Keratol; sides are beautifully embossed and hand painted, tinted gilt scroll work in relief, handle at top fastened with gilt metal ornaments, fine satin lining, fancy gilt metal feet. Size $3\frac{1}{4}\text{x}6\frac{3}{4}\text{x}8\frac{3}{4}$ inches. Doz.............6 00

7670

7671 **Photo holder**, made of fine satin in desirable shades, sides are elegantly struck up and beautifully hand painted, handle at top fastened with dainty gilt metal ornaments, fine satin lining, gilt metal feet. Size $3\frac{1}{4}\text{x}6\frac{3}{4}\text{x}8\frac{3}{4}$ in. Doz...... .. 8 25

1907

114

GELLMAN BROS. MINNEAPOLIS, MINN.

Celluloid "What-not" Ornaments and Novelty Paper Weights

No. 6438 — Celluloid Ornament. Consists of comical dog having painted features, holding a large brightly colored parasol, mounted on black celluloid base. Height, 2⅜ inches; size of base, 1x⅞ inches. Packed each in an individual box and one dozen in carton.

Per doz.... **60¢**

No. 6422—Celluloid Ornament. This is an exact reproduction of a coolie drawing a jinrikisha with a passenger holding a large parasol, mounted on black celluloid base. Height, 1¾ in.; size of base, 1¾x1¼ in. Packed each in an individual box and one dozen in carton.

Per dozen.................. **90¢**

No. 6441—Celluloid Ornament. Consists of two penguins made of black and white celluloid on white celluloid base. Height, 1⅛ inches; size of base, 1½x1⅛ inches. Packed each in an individual box and one dozen in carton.

Per dozen.................. **50¢**

No. 6427 — Celluloid Ornament. Three acrobatic elephants very cleverly made and mounted on each other on all black celluloid base. Height, 1⅞ inches; size of base, 1⅜x1 inch. Packed each in an individual box and one dozen in carton.

Per dozen **60¢**

No. 6444—Celluloid Ornament. Beautifully modeled dog with painted features seated on modernistic celluloid chair. Height, 2¼ inches; width of chair, 1 inch. Packed each in an individual box and one dozen in carton.

Per doz..... **45¢**

No. 6445—Celluloid Ornament. Two parrots very attractively painted in bright colors with painted features, roosting on fancy celluloid chair. Height, 2¼ inches; width of chair, 1⅛ inches. Packed each in an individual box and one dozen in carton.

Per doz. **45¢**

No. 6425—Celluloid Ornament. Consists of perfectly modeled dog with painted features seated near large fancy decorated parasol, mounted on black celluloid base. Height, 1⅞ inches; size of base, 1⅜x¾ inches. Packed each in an individual box and one dozen in carton.

Per doz. **60¢**

No. 6443—Celluloid Ornament. Attractively finished dog with painted features seated on fancy circular celluloid chair. Height, 2½ inches; diameter of chair, 1⅛ inches. Packed each in an individual box and one dozen in carton.

Per dozen **45¢**

No. 6449—Novelty Paper Weight. Celluloid rooster and chicken pecking at several grains of corn on black celluloid base. Complete with propel and repel type mechanical pencil attached by 10-inch silk cord. Height, 2⅜x1⅞ inches. Packed each in box and one dozen in carton.

Per dozen.. **90¢**

No. 6457 — Novelty Paper Weight. Consists of black celluloid base with one large and one baby elephant standing thereon. Has a propel and repel type mechanical pencil attached by 10-inch silk cord. Height, 1¼ inches; size of base, 2⅜x1⅞ inches. Packed each in box and one dozen in carton.

Per dozen........................ **90¢**

No. 6459—Novelty Paper Weight. Consists of two celluloid penguins standing on black celluloid base. The base has a pencil rest and a propel and repel type mechanical pencil attached by a 10-inch silk cord. Height, 1¼ inches; size of base, 2⅜x1⅞ inches. Packed each in box and one dozen in carton.

Per dozen.................................. **90¢**

No. 6458—Novelty Paper Weight. Celluloid dog with painted features seated on black celluloid base with 10-inch silk cord and propel repel type mechanical pencil attached. Height, 1⅝ inches; size of base, 2⅜x1⅞ inches. Packed each in box and one dozen in carton.

Per dozen........................ **90¢**

WHOLESALE PREMIUM MERCHANDISE & NOVELTIES

1936

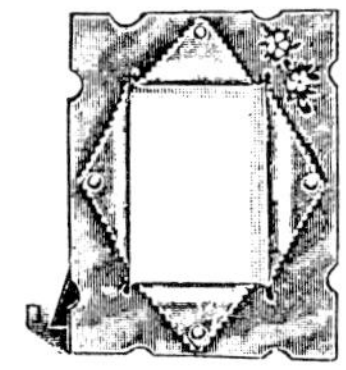

72c. Dozen.

The Solid "Celluloid" Frame.

......A beautiful gem. Size, 6x8 inches. Hand-Painted in corner easel back, assorted colors of celluloid in each dozen.

......Per dozen............................$ 72

1892

TOILET AND MANICURE SETS.

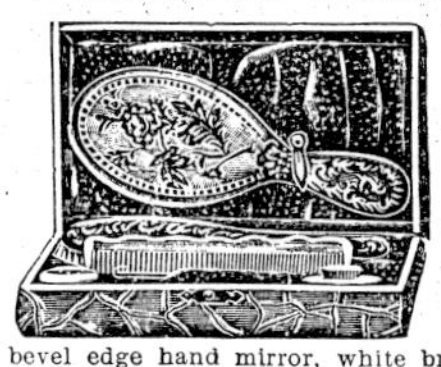

7507 Toilet case, fancy curved and shaped top, size 8⅛x5½x3¾ inches, covered with a **moire silk effect cream color leatherette,** decorated with **gilt** stripes and bouquet of violets in colors, embossed and gilt edges, fancy clasp, sateen lining, large size bevel edge hand mirror, white bristle hair brush to match, also celluloid comb, complete with nail polisher, button hook and salve box.
Doz...9 00

7508 Toilet case, fancy oblong shape with round ends, size 9¾x4¼x3 in. gilt and floral pattern leatherette covered, wide celluloid picture band top, fancy gilt clasp, sateen lining, extra large and handsome bevel mirror with embossed back, white bristle brush and celluloid comb to match. New and very handsome toilet case. Doz..9 00

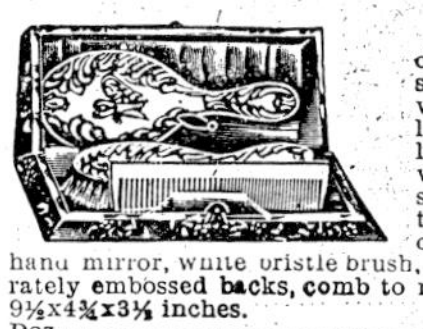

7509 Toilet case, oblong shape with slanting top covered with handsome floral leatherette, broad celluloid picture band with embossed edges, sateen lining, contains 3-piece set including bevel edge hand mirror, white bristle brush, both with very elaborately embossed backs, comb to match, gilt clasp, size 9½x4¾x3⅛ inches.
Doz...9 00

7510 Toilet case, oblong shape, extension base, size 5⅛x10 inches, top covered with wide celluloid picture band with embossed edges, balance covered with **floral** leatherette, fancy moire lining, large size bevel edge embossed back hand mirror, white bristle hair brush, comb to match, gilt clasp. Doz..9 00

7511 Toilet case, size 9½x4¾x3 inches, hinged top, extension case, covered with handsome embossed marble effect paper, celluloid top bearing two three-quarter length portraits artistically executed in colors with flowers in relief, gilt clasp, full sateen lined, bevel edge hand mirror and hair brush with embossed gilt outlined backs, celluloid comb. Doz..............................9 00

7512 Oblong shape new design **toilet case,** size 4x10x2⅝ inches, floral leatherette, covered picture top, gilt clasp and hinges, fancy moire lining. Set includes large size bevel edge embossed back hand mirror, high grade white bristle hair brush, comb and salve box, manicure scissors. Doz..9 00

7513 Toilet case, oblong shape, extension base, size 11x5¼x3 inches high mercerized lined, complete with fine bevel edge hand mirror and high grade white bristle hair brush, each with handsome embossed backs, 6½-inch celluloid comb, case has extension base, hinged top, gilt clasp, top bears beautiful portrait. Doz..............................9 00

7514 Toilet case, square shape with extension base, size 9¾x6½x3 inches, **floral leatherette covered,** very handsome portrait top, contains 6-piece toilet set, consisting of large and heavy bevel edge hand mirror with embossed back, white bristle hair brush to match, comb, manicure scissors, nail polisher, and salve box, fancy gilt clasp. Doz....9 00

7515 Very popular selling **square shape toilet case** with extension base, size 9½x6⅛x2¼ inches, dark red leatherette covered with embossed top, **looks and wears like genuine leather,** with gilt clasp, sateen lining, large and heavy bevel edge hand mirror with elaborately embossed back, white bristle hair brush to match, comb and 3 pieces manicure articles.
Doz..9 50

7516 Toilet case, new fancy design, oblong shape, size 10⅝x5¾ inches wide, full celluloid top with a beautiful floral design in colors, balance covered with a gilt and floral leatherette, handsome gilt clasp, sateen lined, large and heavy bevel edge hand mirror with embossed and beaded back, white bristle hair brush and comb to match. Very attractive new case. Doz..........12 00

7517 Toilet case, oblong shape, slanting top, extension base, size 9½x4¾x3½ inches high, full covered with handsome marbleized celluloid, embossed scroll edges, very handsome floral top, fancy gilt clasp, large size heavy **bevel edge hand mirror** with embossed and gold striped back, extra quality white bristle brush to match, fine and coarse tooth celluloid comb. Very rich and handsome toilet case.
Doz..12 00

7518 Toilet case with oval top, extension base, size 6½x10x3 inches high, tastefully decorated with dainty floral leatherette with floral design on top, hinged back, gilt clasp, fancy mercerized lining. Case contains 6-piece toilet set including full size bevel edge hand mirror with embossed back, extra quality white bristle brush to match, comb, manicure scissors, nail polisher and salve box. Doz......12 00

1936

FANCY CELLULOID TRUMPETS.

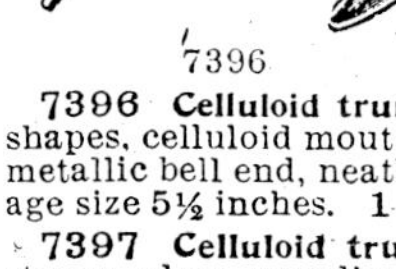

7396

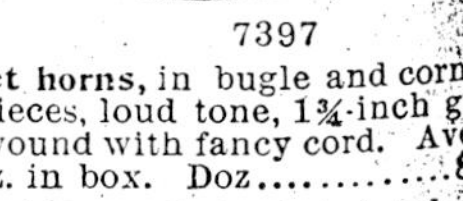

7397

7396 Celluloid trumpet horns, in bugle and cornet shapes, celluloid mouth pieces, loud tone, 1¾-inch gilt metallic bell end, neatly wound with fancy cord. Average size 5½ inches. 1 doz. in box. Doz...............85

7397 Celluloid trumpet horns in assorted colors, strong, clear sounding reed, with assorted figures on outside of horn, average length 6½ inches, complete with ring and cord. Assorted ½ doz. in box. Doz. 2 00

7398

7398 Celluloid Trumpet Horns, assorted in bugle and cornet shapes, wound with chenille cord and tassels, clear sounding trumpet reed, 2½-inch bell shaped end. Size 9 inches. ¼ doz. in box.
Doz.........................2 00

7399 Length 8¼ inches, made of celluloid, finished in pure cream color, horn shape, has heavy metal 3-inch bell end gilt finished, porcelain mouthpiece set in fancy gilt socket, clear sounding clarion reed, single mouthpiece, complete with 24-inch gilt chain. ¼ doz. in box. Doz. 2 00

7399

1907

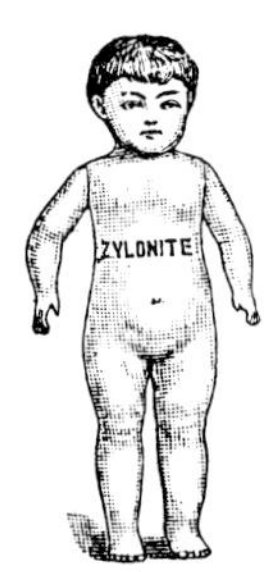

ZYLONITE DOLLS WITH JOINTED ARMS.

Light in wt., practically indestructible, flesh tinted, hollow, well proportioned bodies, good features, painted hair, eyes and eyebrows, jointed arms, free legs.

F2140—3¾ in. 1 doz. in box.............. 85
F2141—6 in. Each in box. 1 doz. in pkg. 1 95
F2142—8 in. Each in box. ⅓ doz. in pkg. 4 10
F2143—11 in. Each in box................ 8 00

1905

FANCY CELLULOID HANDKERCHIEF BOXES.

Each has gilt clasps.

F4131—Imit. birch bark covering in asstd. shades, hinged cover, colored picture in frame, gilt lettering, neatly lined, 6½x5½x1¾. 3 in box. Each, **17c**

F4132—Figured bowknot covering in asstd. tints, chromo picture top, neat glazed lining, 7x6x2½. Each in box. Each, **18c**

4133—Asstd. floral color design tinted covering, colored panel picture top with gilt border, neat puffed figured lining 8x6¼x3½. Each, **33c**

F4134—Asstd. floral designs, tinted ground in asst. shades, colored pictures under celluloid, gilt embossed frame, beveled edge top, puffed mercerized lining, 7¼x6¼x3. Each, **37c**

F4135—Floral gilt and tinted covered, colored pictures under celluloid, mercerized lining, 6¼x6¼x2¾. Each, **50c**

F4136—Asstd. rose, forget me not and carnation floral tinted and bronze covering, colored medallion top, puffed satin lining, 8x6¼x3½. Each **69c**

F4137—Fancy oval front with sunk-in sides, asstd. colored pictures under celluloid, rich floral tinted sides, puffed mercerized lining, 9x7x2¾. Each, **72c**

F4138—Asstd. tints, embossed in scroll and floral designs, fancy lettering in gilt and asstd. tints, puffed satin lining, 7½x7½x3. Each, **95c**

F4139—Fancy wave top, floral decorated gilt covering, colored medallion in gilt embossed frame, rich puffed satin

FANCY CELLULOID JEWEL CASES.

Sizes given are for cases open.

F4115, 18c Each. F4116, 35c Each.

F4115—Asstd. floral design, tinted covering, gilt tracings, chromo picture top, gilt clasp, puffed figured lining, watch holder 5½x5¼x6 ....Each, **18c**

F4116—Imit. quartered oak and mahogany finish, asstd. pictures on cover, gilt clasp, rich puffed mercerized lining, watch holder, 5½x5¼x6 ....Each, **35c**

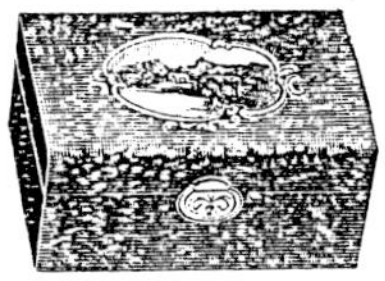

F4117—Marble covering in asstd. shades, colored picture top, puffed satin lining, removable tray, watch holder, velvet ring rack, gilt lock and key, 7½x6¼x7¼. Each, **50c**

F4118—Asstd. colored pictures under celluloid floral tinted sides, gilt lock and key, puffed mercerized lining, watch holder and ring rack, 7½x6x6½. Each, **65c**

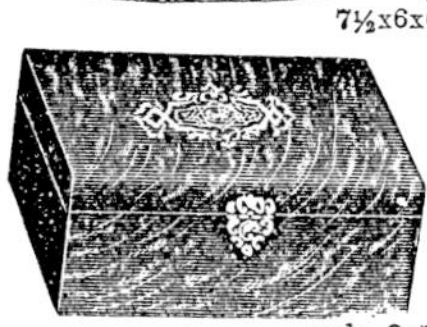

F4119—Imit. quartered oak and mahogany, asstd. scrolls in inlaid effect, gilt clasp, puffed satin lining, watch holder, plush covered ring rack, 8x7x0. Each, **75c**

F4120—Colored picture and floral sprays, under celluloid, embossed borders, rich floral gilt covered sides, gilt clasp, 7¼x5¼ in. mirror in roof, puffed satin lining, watch holder, plush covered ring rack, 8¾x7x8½. Each, **92c**

F4121—Rich floral design covering, colored medallion in gilt embossed frame puffed satin lining, removable tray, watch holder, plush covered ring rack, gilt lock and key, extension drawer, 8x8x9. Each, **$1.25**

F4122—Embossed and colored in imit. stag horn, embossed stag head, puffed satin lining, gilt lock and key, removable tray, watch holder and ring rack, 7½x7x8............Each, **$1.50**

WE ask no favors. In any line all we ask is COMPARISON. We make no claims

1905

ZYLONITE DOLL HEADS.

Painted Hair.

Curly Hair.

At prices now within the reach of all.

Painted Hair—Natural flesh tinted painted, hair, eyebrows and lashes, mouth showing teeth, good features. natural glass eyes. Extremely light in wt. and practically indestructible.

F2144—3½x2⅞. ¼ doz. in box........................ 2 00
F2145—4¼x3½. ¼ doz. in box........................ 3 00
F2146—5⅛x4½. Each in box 4 00

Curly Hair—As above with fine quality sewed wigs, curly hair dress. Each in box.

F2147—4⅛x3⅛............... 4 10
F2148—5½x4¼............... 5 90
F2149—6¼x5................ 8 00

1905

Celluloid Gift Bookmarks

Bookmarks made of celluloid, beautifully decorated and finished in colors. Practical as well as attractive. The styles here shown are shaped and perforated to receive the edge of the leaf, thus securing their position. Each Bookmark is 4⅞ inches long. Five styles, as follows:

No. 4898-E. Dove Design.—With 23d Psalm.

No. 4896-E. Lilies and Roses.—With Beatitudes upon it.

No. 4893-E. Mothers.—Design of mother and babe, with encouraging sentiments.

No. 4392-E. Thanksgiving.—Decorated with turkey and thanksgiving sentiments.

No. 4895-E. Pansies.—With the Ten Commandments.

Price, 5 cents each. Three or more, **4 cents each,** assorted as desired, postpaid.

Heart Bookmark

No. 4899-E.—White sheet celluloid bookmark, heart-shaped, with double silk cord and tassel. Decorated in colors, with choice Scripture Text on face, and words, "With best wishes," on the back. Size, 2 x 2¼ inches. Cord eight inches. Five or more, **3 cents each, postpaid.**

Paper Cutter

No. 4900-E.—Paper cutter and bookmark Solid celluloid, decorated in colors on both sides. Scripture verse. "Serve the Lord with gladness," and the word "Greetings" daintily printed on each. Size, ¾ x 3⅜. Five or more, **3 cents each,** postpaid.

1917

CHRISTMAS SOUVENIR PIN.

Celluloid button pins, engraved designs. Dainty souvenirs. Usually distributed at the Christmas entertainment on entering church. Four styles:

No. 2540-D. Christmas Star Souvenir Button.—In colors, showing Christmas Star and gate of Bethlehem.

No. 2561-D. Christmas Holly Button Pin.—Orange yellow pin with holly decorations in red and green.

No. 1616-D. Santa Claus Souvenir Pin.—With picture of Santa Claus. Beautifully printed in bright colors.

No. 1660-D. Christmas Tree Souvenir Button.—Gorgeous pin with Christmas tree in colors.

Ten or more, either kind, **1 cent** each, postpaid. With celluloid bows attached, **2 cents each.**

1917

CELLULOID ROLY POLY FOR THE BABY

No. J842. Designed to please the babies. The main attraction of this Roly Poly is that when it rolls over it will spring back up by itself. It is made of brightly colored celluloid that will catch the baby's eye. Ship. wt.. 1¾ oz. SELLING PRICE 10c to 15c

Sample 11c; Dozen $1.18
2 Dozen $2.19

1952

56 DAVID C. COOK PUBLISHING COMPANY'S ANNUAL CATALOGUE FOR THE YEAR 1917.

"Working" Class Celluloid Button Pins

Beautiful celluloid class button pins, ready-made and carried in stock for the following new and attractive class names. More than 1,000,000 sold.

Prices, all pins shown on this page, ten or more, **1 cent each**, postpaid.

No. 2409-B. Class Names and Pins.—Booklet giving more than one hundred class names, with suggestions for mottoes, aims, sentiments and colors. Also cuts of class pins suitable for each one. **Price, 3 cents**, postpaid.

We will send sample of any four pins shown here, and booklet No. 2409-B, telling all about class plans, etc., for each class for which these pins are designed, for **5 cents**, postpaid.

No. 1600-D.—The Its; or "The I. T.'s," or "The We're It's." Purple button, gold lettering. For classes of young or old. Object: Growth through effort. Motto, "I'll try;" or "Strive for the mastery."

No. 1601-D. — True Blue; or, "The Reliables." White lettering trimmed with gold, on blue button. Class object, Personal honor, reliability, loyalty, etc.

No. 1602-D. — The White Tigers; or, "The S. & T.'s" (Strong and True), or "Th Anti-Yellows." Pure white tiger on gold button, red lettering. Class motto. "Aggressive, but not yellow."

No. 1603-D. — The Gang; or, "The G. & G. Class." Gray button, red and blue lettering, trimmed with gold. Novel class button; sure to please wonderfully. Motto, "Grace and grit."

No. 1604-D.—The Anti-can'ts; or, "Invincibles." Curiously wrought design. Simple, and yet not easily read until interpreted. Buff button, lettering in blue and red. Object, To overcome the tendency to say, "I can't," etc.

No. 1605-D.—Cheerful Wigglers; or, "Best-You-Can-Do Club." Green button, red lettering, pink wiggler. Motto, "A merry heart doeth good like a medicine."

No. 1606-D. — The Diggers; or, "Treasure Seekers," or "Persistency Club." Dark gold button, silver pick, green lettering. Object, To encourage persistency, etc.

No. 1609-D.—The Tri-Mu Class; or, "The Three M.'s" (Muscle, Mind, Morals). Blue button, gold triangle and letters, figure in red. Object, to rightly develop the three sides of our nature.

No. 2526-D. Comrades; or, The Comrade Class.—Dark red pin with hands clasped. Letters in gold. Motto, "We're Comrades." Aim, To develop the fraternal spirit.

No. 2527-D. Winners.—Light brown pin, with laurel wreath in green. Letters blue. Motto, "On to the goal."

No. 2594-D. Busy Bees.—Ground of pin in brown and blue. Gold bee, lettering white. Motto: "Always abounding." "We gather honey all the day," etc. Aim: To make life sweeter: To be both busy and happy. Colors: Brown and gold.

No. 2591-D. Sunbeam Class. — Blue ground, orange-yellow sunbeams. Motto, "Lights in the World," "Be Cheerful." Aim, Scatter sunshine. Colors, blue, red and yellow.

No. 2593-D. Willing Workers.—Royal blue button, white letters. Class mottoes, "Cultivating the 'We will' spirit," "Working Willingly," "Serving Gladly," "Where there's a will, there's a way." Aims, "To do with our might what our hands find to do," "To work willingly," "To will to work and to work with a will."

No. 1613-D. Missionary Class.—Letters "C. T. N.," standing for words, "Carry the News." Blue button, white flag, gold lettering. For class having as its object to help in missionary work.

No. 2528-D. Live Wires.—Navy blue button pin with yellow, red and green setting. Motto, "Alive to every opportunity. Aim, To be alert and energetic in all class relations.

No. 2529-D. Sunshine; or, Sunshine Class.—Green and gold pin with class name in white letters. Motto, "To carry sunshine everywhere." Class song, "Scatter Sunshine."

No. 2530-D. Leaders.—Light blue pin with pennant in dark blue, lettering in yellow. Motto, "To lead in every good work."

No. 2531-D. King's Daughters. — Light brown pin with crown in gold and blue. Motto, "To be loyal in thought, word and deed." Aim, To do all in His Name. Class song, "I am the child of a King."

No. 1627-D. Twilight Club.—Large white star in blue sky, church spire in distance. Lettering in orange. Suitable for class of young men or young women whose object may be regular attendance at evening church service, young people's society, or some evening meeting of their own.

No. 1630-D. The S. P. I. Class.—Lettering in white, on wide stripes of purple, red and orange. The letters stand for "Spiritual, Physical and Intellectual." Aim of class to cultivate these ideals.

No. 1622-D. Young Men's Bible Class Pin.—For classes of young men. Gold ground button pin with white shield, deep blue cross and lettering.

No. 1623-D. Young Ladies' Bible Class Pin.—For classes of young ladies. Silver ground button pin with white shield, light blue band across same, and light blue lettering.

No. 1624-D. Bible Class.—Blue button pin with white letters, "B. C." as per cut opposite. Suitable for any adult Bible class—old or young, men, women or mixed.

No. 1631-D. The Goldbugs.—Dark blue ground, bugs in yellow. For classes having as their object the gathering of true riches.

No. 2521-D. The Swastika Class, or Good Fortune Class, or Lucky Boys, or Lucky Girls. Motto: Pluck brings luck, etc.

No. 1669-D. Boys' Brotherhood.—Or B. B's. Blue ground with gold and white monogram. Motto: "Friendliness." Aim: To show a friendly brotherhood spirit, and to seek to win others to the Friend of friends.

No. 1667-D. Upstreamers; or, The U S. Class.—Light blue button with yellow and green foreground. Lettering in blue and gold. Boat in red and white. For young men's classes. Motto: "Pulling together." Colors: blue for truth, gold for genuineness. Aim: to work, not drift.

No. 1668-D. The Stars.—Blue ground button with star in silver. Lettering in white and red. Class motto: "Hitch your wagon to a star." "Shining." Aim: To keep high ideals and seek to attain them. Colors: blue for honor, silver for purity, red for courage and fellowship.

No. 1632-D. The Class in the Corner; or, The C. C's.—Orange yellow ground, green laurel wreath and letters, white band. As the name indicates, for the class in the corner. Motto: "In a corner, but not cornered." "Help the fellow that is cornered," "To brighten dark corners," etc.

No. 1671-D. Knights of Honor.—Button in red, Shield in gold, cross in white, lettering in deep blue. Motto: "Do what's right, come what may." Aim: To be chivalrous in our dealings with others.

No. 2513-D. The Two "T's"; or, The Tried and True Class.—Solid blue button, lettering in white. Motto: Through all trying experiences be true.

No. 2514-D. Brotherhood; or, Brotherhood Class. — Blue button with red shield and white cross. Suitable for all.

No. 2517-D. The Friendly Bible Class, or The F. B. C. Class.—Blue button pin, lettering in white. Purpose: To cultivate the spirit of friendship.

No. 2518-D. The Who Do Class.—Black ground pin with lettering in white. Purpose: To encourage doing things as a class, and to do away with the "hoodoos" of discouragement.

No. 2519-D. Beacon Lights. — Red button pin, with design of lighthouse and lettering in white. Motto: Let your light shine.

No. 2520-D. Golden Hours Class. — Red ground with hourglass in gold and white, lettering in blue. Motto: Make the hours golden. Bible text: "Redeeming the time."

No. 2544-D Merry Maids.—Unique pin with the initials M. M. in gold on white ground. Border red and blue. For Young Ladies' Classes. Motto, "To help others we must learn to be happy." Object, To be joyous and scatter sunshine. Bible Text, Prov. 17: 22.

No 2545-D. Mountain Boys.—Terra Cotta ground, Mountain and sky in gray, blue and white. Lettering orange. Class motto: "We seek the heights and help others climb." Scripture verse, Ps. 121: 1.

No. 2546-D. Champions.—Buff ground. Design in terra cotta, brown and white, lettering blue. Class motto, "Dare and do," or Champions of the right," etc.

No. 2547-D. Ever Ready Class.—Yellow ground, lettering in blue. Motto, "Improving opportunities," or "On the Job," "Up and doing," etc.

No. 2548-D. Royal Oaks.—Terra cotta ground, design of oak tree in natural colors, lettering royal blue. Motto, "To encourage bravery, hospitality, sturdiness, reliability, moral strength."

No. 2549-D. Choir Loft Class.—Purple ground, organ pipes in terra cotta, lettering orange. Class object: Making the most of circumstances. Motto, "Surmount obstacles."

No. 2550-D. Aeronauts.—Sky-blue ground, design in white, lettering, orange. Class object, To rise above circumstances. To aspire. Motto, "Where the outlook is not good, try the uplift."

No. 2551-D. Black Cats. — Red ground, head of cat in black, lettering in white. Object, Mousing around. Scripture verse: Romans 12: 9.

No. 2553-D. Lotus Class.—Blue ground, Lotus Flower (or Water Rose) in white and green. Lettering orange. Young Ladies' Class emblem. Motto, "Remembering the good of the past, and appreciating the blessings of the present," or "Determining to be contented and seeking to make others happy."

No. 1619-D. Baraca Pin.—For classes organized or about to start under Baraca class plans. In Baraca colors, royal blue and white.

No. 1620-D. Philathea Pin.—For classes organized or about to organize in accordance with Philathea plans. Philathea colors, light blue and white. White crescent, silver trimmings.

Price.—All celluloid Class Button Pins shown on this page, **1 cent each,** when ordered in lots of ten or more.

Pins with Bows.—(See cut below.) Bows of pressed sheet celluloid in imitation of ribbon. Very pretty. Furnished with all class pins on this page where desired. Price of pins with bows, **2 cents each, postpaid.**

Emblematic Class Pins.—For great number of class names. See page 50.

PINS MADE TO ORDER.

Celluloid Button Pins made to order to suit any class at very low prices. See page 21.

See also **Metal Pins,** made to order, on page 21.

INDIVIDUAL CLASS BOOKLETS.

Booklets provided, one each for class name as per pins shown above, except young ladies' class and young men's. Give interesting class history, and practical suggestions for making the most of the name, etc. **Price, 2 cents each.** Sample of any three class pins, with booklets to match, for **5 cents,** postpaid.

These booklets aid greatly in choosing a most desirable name as well as in making the most of it later.

CLASS BADGES.

Felt badges, pennant-shaped, size, 4 x 1½, furnished for all class names as per pins shown on this page. In lots of ten or more, **1 cent each,** postpaid.

1917

WISCONSIN DELUXE COMPANY 191

FUR AND CHENILLE MONKEYS—CELLULOID FEATHER DOLLS

No. 4918 — **Celluloid Feather Doll with Hat.** With gold painted hair, and silver mica decoration. Full facial painted features. Bonnet and skirt made of red and blue feathers. Overall height 4 in. Packed 3 dozen to box.
Per Dozen $ .85
Per Gross **$8.00**

No. 4961—Same description as listed above, but larger. Overall height 5 in.
Per Dozen $1.00
Per Gross **$9.35**

No. 4955—Same description as listed above, but larger. Overall height 6 in. Packed 2 dozen to box.
Per Dozen $1.10
Per Gross **$10.00**

No. 4930—Same description as listed above, but larger. Overall height 7 in. Packed 1 dozen to box.
Per Dozen $1.70
Per Gross **$17.00**

No. 4956—Same description as listed above, but larger. Overall height 8 in. Packed 2 dozen to box.
Per Dozen $2.25
Per Gross **$24.00**

No. 4932—Same description as listed above, but larger. Overall height 9 in. Packed 1 dozen to box.
Per Dozen $2.65
Per Gross **$28.00**

No. 4971—Same description as listed above, but larger. Overall height 10 in. 1 dozen to box.
Per Dozen $3.20
Per Gross **$33.60**

No. 4959—Same description as listed above, but larger. Overall height 12 in. Packed 1 dozen to box.
Per Dozen $3.80
Per Gross **$42.00**

No. 4948—**Fur Monkey.** This jumping monkey comes with high hat and spring body fully covered with fur. Overall height 4 in. Imported. Packed 2 dozen to box.
Per Dozen $ .50
Per Gross **$4.00**

No. 4949—Same description as listed above, but in a 5 in. monkey. Packed 2 dozen to box.
Per Dozen $ .60
Per Gross **$4.70**

No. 4942—**Fur Monkey.** Has plastic head and Hi-hat. Full painted features. Body made of fur, attached to brass spring. A novel action toy, overall length 6 in. Imported.
Per Dozen $ .75
Per Gross **$7.00**

No. 4990—Same description as listed above, but larger. Overall height 7 in. Packed 2 dozen to box.
Per Dozen $ .85
Per Gross **$8.30**

No. 4933—Same description as listed above, but larger. Overall height 8 in. Packed 1 dozen to box.
Per Dozen $ .95
Per Gross **$9.00**

No. 4980—Same description as listed above, but larger. Overall height 9 in. Packed 1 dozen to box.
Per Dozen $1.40
Per Gross **$13.00**

No. 4989—Same description as listed above, but larger. Overall height 10 in. With Felt Hat.
Per Dozen $1.60
Per Gross **$15.00**

No. 4950—Same description as listed above, but larger. Overall height 12 in. Packed 1 dozen to box.
Per Dozen $3.00
Per Gross **$33.60**

No. 4964—**Hula Skirt Celluloid Feather Doll.** A new number in celluloid feather dolls—Real flesh. Has rayon silk fringed hula skirt, celluloid cane, bead necklace and metallic paper clown hat for trim. Size 5 in. Packed 3 doz. to box.
Per Dozen $1.00
Per Gross **$9.25**

No. 4977—**Hula Skirt Celluloid Feather Doll.** Same general description as No. 4964 with the exception of having metallic paper plug hat instead of clown hat, and 7 in. in height. Packed 1 dozen.
Per Dozen $1.80
Per Gross **$18.00**

No. 4983—**Hula Skirt Celluloid Feather Doll.** Same general description as No. 4977 but larger, size 9 in. Packed 1 dozen to box.
Per Dozen $2.75
Per Gross **$28.70**

No. 4920—**Bathing Girl.** Gold painted hair; full painted facial features. Arms on elastic, movable. Imported. Dressed in painted bathing suit. Height 6½ in. Packed 1 dozen to box.
Per Dozen $1.75
Per Gross **$18.00**

No. 4976—**Feather Dresses.** In assorted color feathers made to fit dolls 6 inches high.
Per Dozen $1.00
Per Gross **$10.00**

No. 4963—**Feather Dresses.** In assorted color feathers made to fit dolls 8 to 9 inches high.
Per Dozen $1.25
Per Gross **$12.80**

No. 4945—**Fez Hat Fur Monkey.** A unique new number. Fur body with celluloid head and fez hat, hands and feet. With ornamental pipe. Height 7 in. Packed 2 dozen to box.
Per Dozen $1.00
Per Gross **$9.35**

No. 4951—**Fez Hat Fur Monkey.** Same general description as No. 4945 listed above but larger, size 10 in. Packed 1 dozen to box.
Per Dozen $2.40
Per Gross **$22.70**

No. 4957—**Fez Hat Fur Monkey.** Same general description as No. 4945 listed above but larger, size 12 in. Packed 1 dozen to box.
Per Dozen $3.25
Per Gross **$35.30**

1952

AMUSING FLOATS FOR BATH TUB PLAY

Highly Colored Ivorette Toys

6 for 42¢

All these toys float in an upright position. Watch baby smile when bath-time comes with these! Our assortment is beautiful with its two pretty swans measuring 3¼ inches long, one 3⅛-in. goose, one 3⅛-in. duck, one 2¾-in. green frog, and one 4½-in. gold fish. The colors are fast and won't come off.

48 G 2787—Postpaid. 6 for 42¢

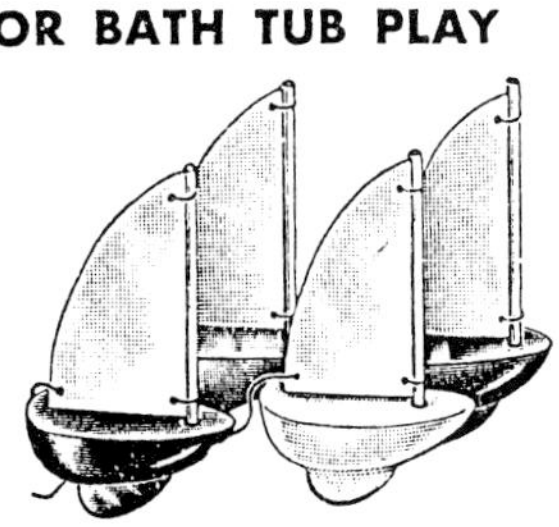

Bobby Bath Boat Fleet

Bright Colors

4 for 45¢

Four little Vessels that actually sail. Each decorated in a clear vivid color—yellow, blue, red and green. Boats are about 3¼ inches in length. Height from top of mast to bottom of keel 5¾ inches. The sails are made out of very fine ivorette. Mast and hull are wood, with the keel of steel.

48 G 2786—Postpaid. 4 for 45¢

1930

10 INCH FEATHERED CELLULOID DOLL

No. J355. A typical little carnival lady—jointed arms, painted big blue eyes and golden hair. Wears a halo of bright colored feathers to match her perky little feathered skirt. Ship. wt. 1½ oz.
SUGGESTED SELLING PRICE 49c
Sample 33c; Dozen $3.68
10 Dozen $34.80

No. J356. Same as above, stands 9" tall. Ship. wt. 1 oz.
SUGGESTED SELLING PRICE 39c
Sample 27c; Dozen $2.98
10 Dozen $28.60

No. J351. 5" tall. Ship. wt. ½ oz.
SELLING PRICE 15c
Sample 9c; Doz. 98c; 10 Doz. $8.90

No. J353. 7" tall. Ship. wt. ¾ oz.
SELLING PRICE 29c
Sample 18c; Dozen $1.98.
10 Dozen $18.90

1952

1905

CELLULOID BABY DOLLS.

A new, indestructible line, made of celluloid; featherweight; very attractive features. This is the first time that this line has been cheap enough for us to handle.

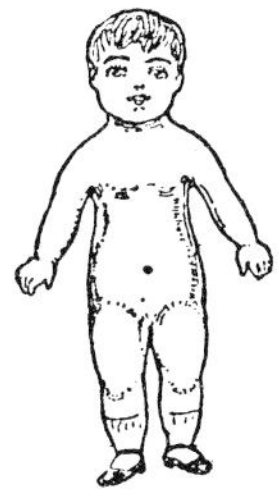

D 2563—As described above; 4¾ inches long	1.90
D 2564—Stout bodies; 6¼ inches long	3.95
D 2565—Extra stout, but very light weight; made of durable celluloid; 8¼ inches long..	6.00
D 2566—Large celluloid baby dolls; 9½ inches long.........	7.50

1931

HAND PAINTED ZYLONITE RATTLE ASSORTMENT

12 big selling styles in a practical partitioned display box. Order an assortment now

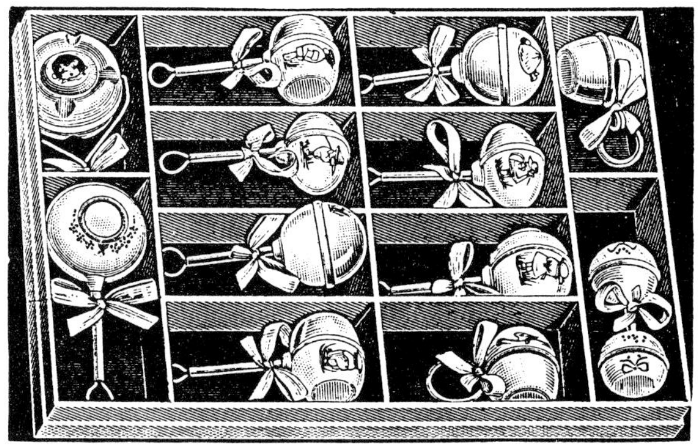

F4476—12 styles, 5¼ to 6½ in., latest novelty shapes, well made, each with finest enamel hand decorations in pastel tints—forget-me-nots, blue birds, storks, chicks, infants, "Baby," etc., large pink or blue bows, some with teething rings, others with loops, solid and combination pink, blue and white. Asstd. 1 doz. in partitioned box. Doz. **$2.15**

IMPORTED CELLULOID ROLY POLY DOLLS

F9461—6 styles, 3¼ in., juvenile characters, with embossed toys in hand, bright colored costumes, weighted bottom makes resume upright position when pushed over. Assorted 2 doz. in box. (Total $1.20) Doz. **60c**

Above: 1920

Left: 1923

Below: 1905

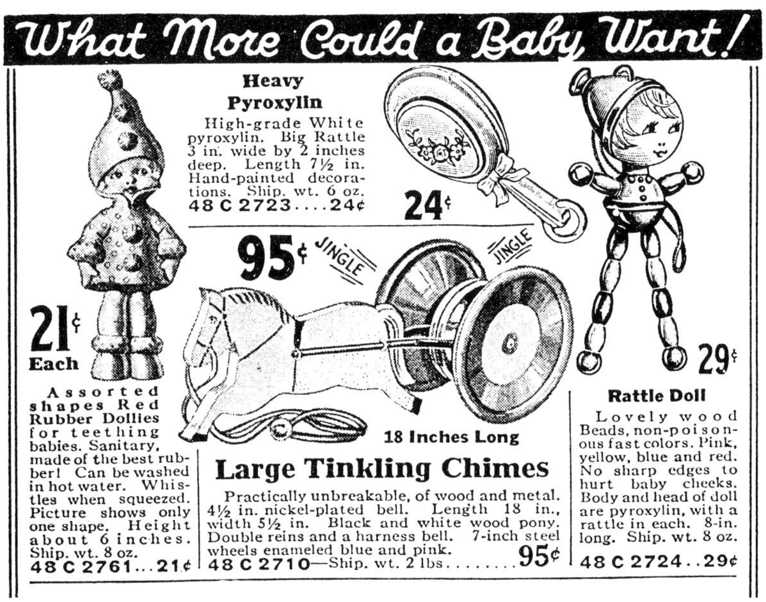

1931

NOVELTY CELLULOID RATTLES.

Per Doz.

D 1892. $2.25.

D 1892—An assortment of French celluloid rattles, beautifully modeled laughing baby faces and twisted celluloid handles. These are not only very attractive, but are very durable, light weight and harmless 2.25

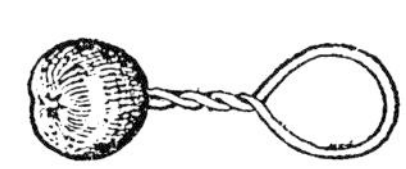

D 1893. $2.25 Doz.

D 1893—An assortment of beautiful celluloid rattles; twisted celluloid handles, representing red cheeked apples, pears, oranges, etc., tinted in natural colors. Both of these numbers are amongst the most pleasing we have ever shown in this department........ 2.25

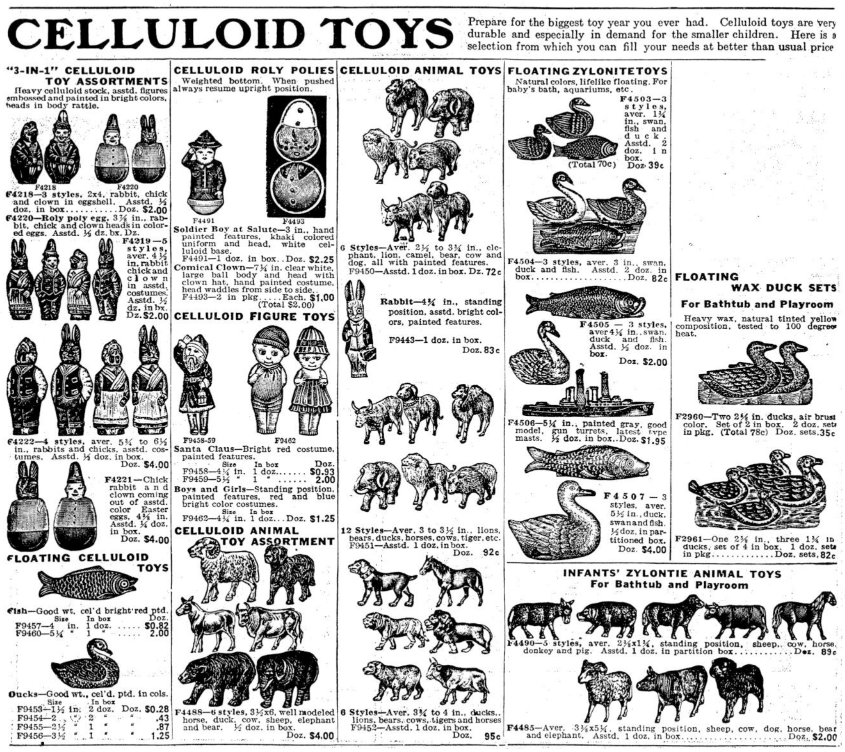

1920

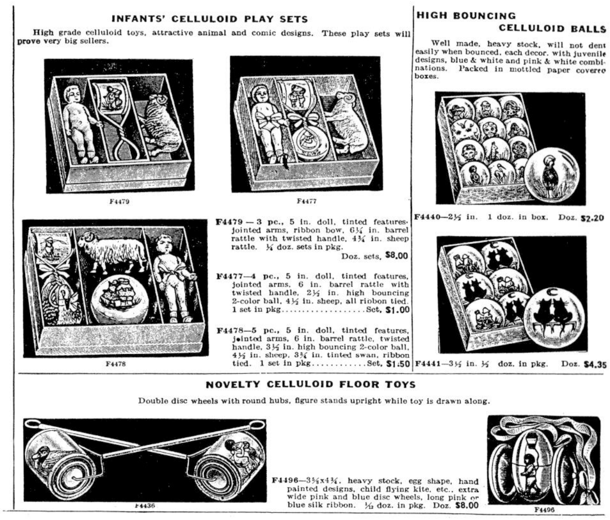

1920

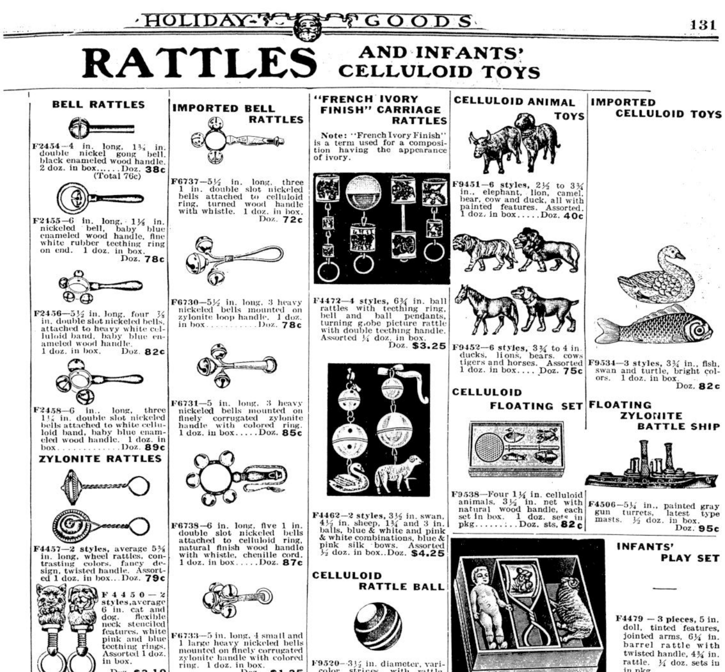

1923

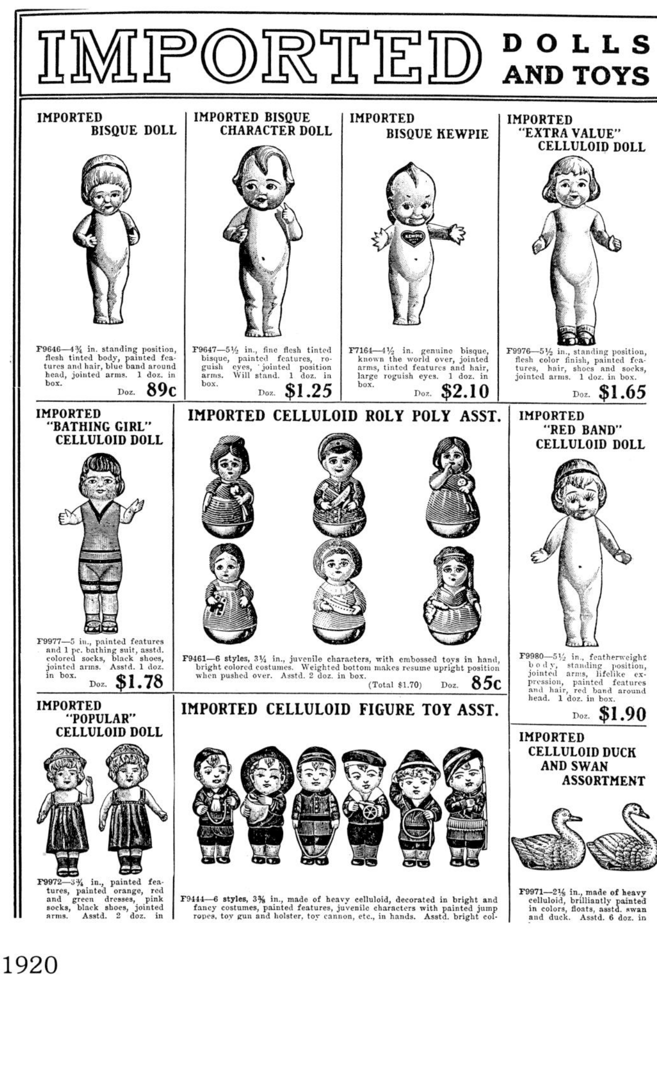

1920

IMPORTED CELLULOID DOLLS

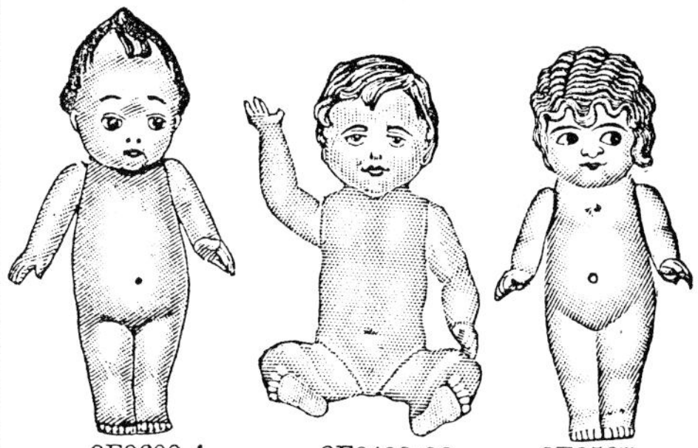

2F9602-4 2F9428-66 2F9527

Jointed arms, painted features and hair.

2F9602—2½ in. 3 doz. in box. Doz **32c** Gro **$3.70**

2F9603—3½ in. 2 doz. in box. Doz T. O.

2F9604—5½ in. 1 doz. in box.. Doz T. O. Gro

2F9430-36 2F9541

Jointed arms, painted features and hair, red band around head.

2F9430—3½ in. high. 2 doz. in box. Gro **$4.00** Doz **35c**

2F9434—6 in. high. 1 doz. in box. Doz **85c** Gro **$9.80**

2F9436—8 in. high. ½ doz. in box. Doz **$1.95**

Santa Claus, painted features, costume and shoes. 1 doz. in box.

2F9541 — 5 in. high. Gro Doz T. O.

Variety Celluloid Assortments

Painted features and costumes.

2F9601—12 style novelties, 2¾ in. high, painted features and costumes, boys and girls. Asstd. 3 doz. in box. Doz T. O. Gro

2F9531—6 styles, some with jointed arms, others with jointed arms and

Jointed arms and legs, painted features and hair

2F9428—4 in. high. 2 doz. in box. Doz T. O. Gro

2F9429—5½ in. high. 1 doz. in box.. Doz **89c** Gro **$9.20**

2F9466—8 in. high. ½ doz. in box. Doz **$2.10**

Jointed arms, painted features and hair.

2F9527—6 in. high. 1 doz. in box.. Doz T. O. Gro

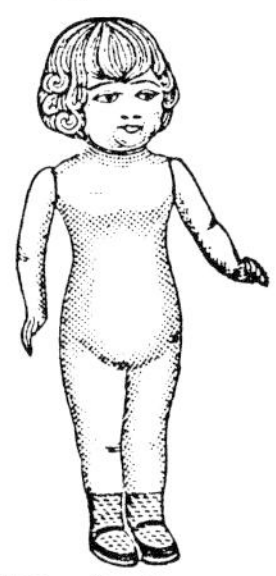

2F9522—6 in., painted features and hair, shoes and stockings, jointed arms. 1 doz. in box Doz **87c** Gro **$9.85**

AMERICAN CELLULOID DOLLS

Painted features, hollow featherweight bodies with thick celluloid shells.

Baby Models — Painted hair, jointed arms and legs, stout bodies, elastic strung.

1F5015—3¼ in. high. 1 doz. in box **84c**

1F5016—5¼ in. high, ½ doz. in box. Doz **$1.90**

1F5017 — 8 in. high. ½ doz. in box. Doz **$4.25**

1F5018—10 in. high ½ doz. in box. Doz **$8.40**

Baby Model — Mohair wig, jointed arms and legs, stout body, elastic strung.

1F5020—6¼ in. high. ¼ doz. in box.

1925

"FRENCH IVORY FINISH" BUFFERS—With Boat and Removable Chamois

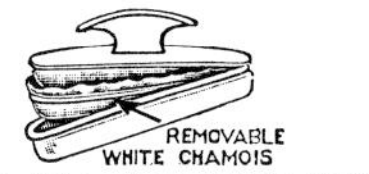

J1160—4¾ in. long, grained stock. ½ doz. in box....... Doz. **$4.20**

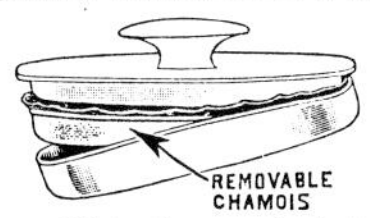

J1161—6¼ in. long, grained stock. ½ doz. in box....... Doz. **$7.00**

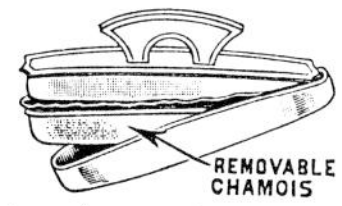

J1162—5¼ in. long, fancy shape, heavy grained stock. 1 in box. Each, **85c**

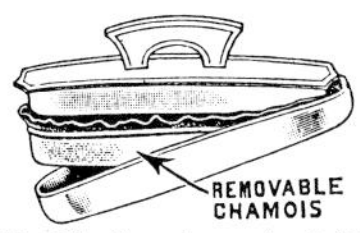

J1163—6 in. long, fancy shape, heavy grained stock. 1 in box. Each, **$1.10**

"FRENCH IVORY FINISH" PERFUME BOTTLE HOLDERS

Grained stock, flanged base.

J1132—3½ in. high, 1½ in. wide, round, crystal bottle, embossed stopper. 1 doz. in box. Doz. **$2.25**

J1133—3 styles, aver. 4 in. high, 2 in. wide, square, round and hexagon crystal bottles, frosted and plain glass stoppers. Asstd. ½ doz. in box............. Doz. **$4.00**

J1134—3 styles, aver. 4¼ in. high, 2 in. wide, square, oval and hexagon crystal bottles, heavy embossed frosted glass stoppers. Asstd. ½ doz. in box......... Doz. **$6.00**

J1135—4 in. high, 3½ in wide, 2 bottles with embossed frosted glass stoppers. 1 in box... Each, **95c**

CUTICLE SCISSORS

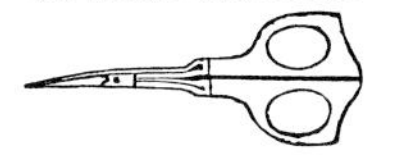

J1131—4⅞ in., polished nickel plated, tempered steel, large fancy shape handles. ½ doz. in pkg. Doz. **$8.00**

SOAP BOX

J1130—3⅜x2¾ in., heavy grained stock, fancy shape. ½ doz. in box. Doz. **$3.95**

"FRENCH IVORY FINISH" JEWEL BOXES

J1138—4⅛x3½x1½ in., grained stock, hinged top, fancy shape, beveled edge, puff lined. 1 in box. Each. **95c**

J1140—5x3¾x1¾ in., heavy grained stock, fancy beveled edge cover, velvet lined, hinged cover. 1 in box........ Each, **$1.75**

COMBINATION PIN CUSHION AND JEWEL BOX

J1167—3½x2¼ in., fancy shape, beveled edge, ribbed bands, removable velvet pin cushion. 1 in box. Each, **85c**

1923

"FRENCH IVORY FINISH" CANDLESTICKS

Grained stock, complete with white enameled metal socket, candle shade and holder.

J1168—11 in. high over all, 5½ in. round candlestick with ribbed bands, blue fluted linen shade with silk braid trim. 1 in box. Each, **$1.10**

J1169—11½ in. high over all, 6½ in. round candlestick with ribbed bands, 4¼ in. base, rose fluted linen shade with silk braid trim. 1 in box......... Each, **$1.85**

1923

INFANTS' TOILET SETS "FRENCH IVORY FINISH"

"French Ivory Finish" is a term used for a composition material having the appearance of ivory

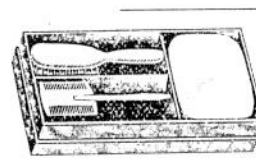

J1111—3 pcs., fine tooth comb with handle, hair brush, soap box, each set in box. ½ doz. sets in carton. Doz. sets, **$4.20**

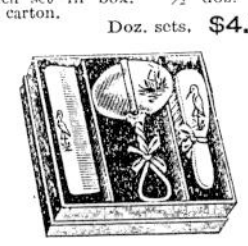

J1102—3 pcs., hand painted stork design, 4¾ in. comb, silk ribbon tied hair brush and 2 color egg shape rattle. ½ doz. sets in box........ Doz. sets, **$7.00**

J1112—3 pcs., hand painted juvenile design, fine tooth comb with handle, hair brush, 2 color rattle, all silk ribbon tied. ½ doz. sets in box. Doz. sets, **$9.00**

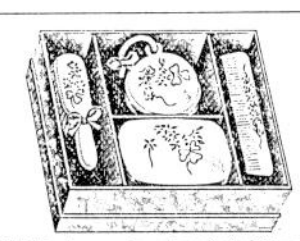

J1104—4 pcs., hand painted forget-me-not decoration hair brush, 4¾ in. comb, ribbon tied, blue and white rattle, soap box, ribbon tied. 1 set in box. SET (4 pcs.) **78c**

J1113—4 pcs., hand painted rose and forget-me-not decoration, 4¾ in. comb, soap box, ribbon tied hair brush and rubber sponge. 1 set in box. SET (4 pcs.) **95c**

J1106—4 pcs., hand painted pink and yellow juvenile decoration, soap box, puff box, ribbon tied hair brush, fine tooth comb. 1 set in box. SET (4 pcs.) **$1.00**

J1114—5 pcs., hand painted floral design, fine tooth comb with handle, hair brush, soap box, rubber sponge, celluloid doll, ribbon tied. 1 set in box. SET (5 pcs.) **$1.20**

J1109—6 pcs., hand painted blue bird and rose decoration, soap box, puff box, ribbon tied hair brush, fine tooth comb, blue and white rattle with teething ring, blue silk powder puff. 1 set in box. SET (6 pcs.) **$1.50**

J1115—5 pcs., hand painted stork, hair brush, 4¾ in. comb, rattle with teething ring, soap box, puff box. Each set in puff lined box. SET (5 pcs.) **$1.60**

A Special Selection of Infants' Toilet Sets "French Ivory Finish"

Very attractive, hand painted designs. The most active sellers, specially priced in one assortment.

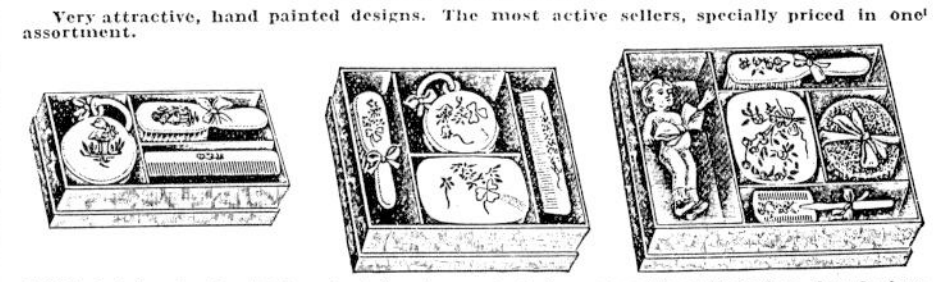

J1117—6 styles, hand painted rosebud, forget-me-not, stork, swan and juvenile designs, four 3 piece and one 4 piece and one 5 piece set, all with solid back soft white hair brushes. Each set in partitioned box. 6 sets in carton. (Total $4.50) Set, **75c**

1923

CELLULOID DOLLS

Heavy celluloid bodies, painted features and hair unless stated, hollow featherweight bodies.

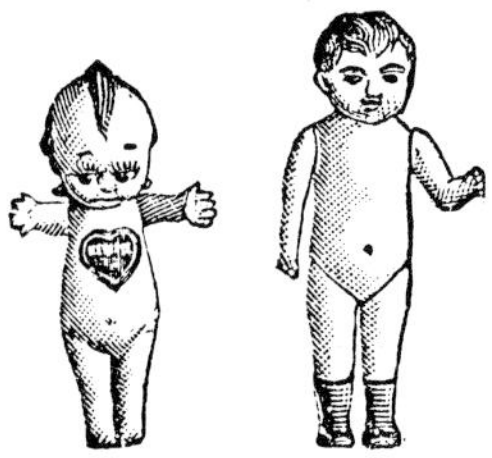

F7808 F7163

Kewpies—Jointed arms, too well known to need description.

F7808—3 in. 1 doz. in box. Doz. **72c**

Floating Dolls—Fixed legs.
F7163—4¾ in., fixed arms. 1 doz. in box....Doz. **45c**

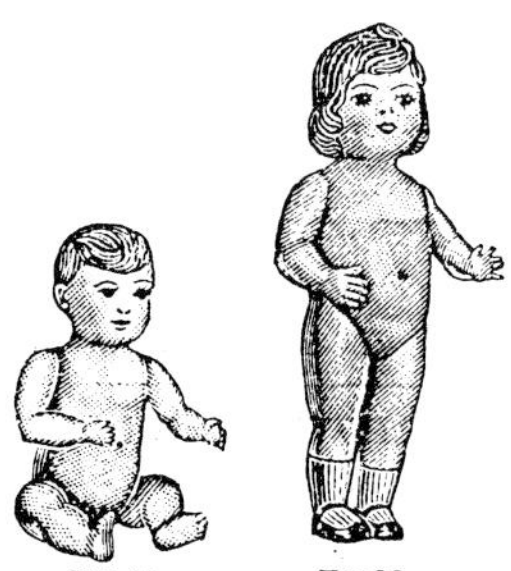

F7166 F7188

Baby Model—Jointed arms and legs, stout bodies, elastic strung, painted hair.

F7166—3¼ in. high. 1 doz. in box.........Doz. **82c**
F7167—5¼ in. high. ½ doz. in box.......Doz. **$2.00**
F7168—8 in. high. ½ doz. in box.......Doz. **$4.40**
F7169—10 in. ½ doz. in box.........Doz. **$7.25**

Upright Dolls—Moving arms and legs, painted shoes and stockings.

F7188—5¾ in. high. ½ doz. in box.......Doz. **$2.25**
F7189—7½ in. high. ½ doz. in box.......Doz. **$4.00**

Baby Model—Jointed arms and legs, **mohair wig,** strong body, elastic strung.

F7185—5¼ in. high. ¼ doz. in box.......Doz. **$2.15**
F7186—6¼ in. high. ½ doz. in box.......Doz. **$4.40**

IMPORTED CELLULOID DOLLS

Red Band Baby Dolls—Standing position, jointed arms, lifelike expression, painted features and hair, red band around head.

	Ht.	In box.	Doz.
F9430—	3½ in.	2 doz.	**$0.45**
F9431—	4 "	1 "	**.60**
F9434—	6 "	½ "	**1.15**
F9436—	8 "	½ "	**1.95**

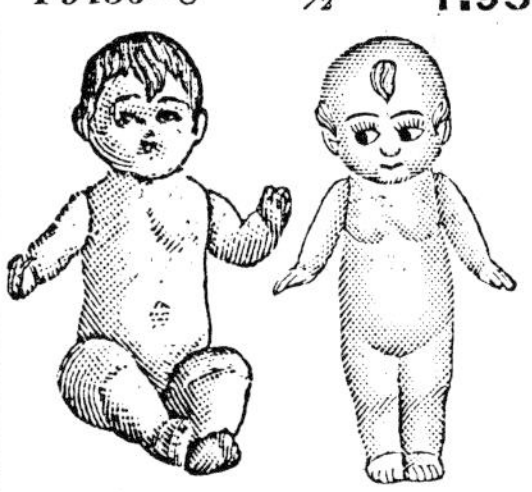

F9428 F9523

Jointed Arms and Legs—Sitting position, painted features and hair.

	Ht.	In box.	Doz.
F9428—	3¾ in.	2 doz.	**$0.65**
F9429—	5 "	1 "	**.85**
F9466—	8 "	½ "	**2.10**

F9523—4¾ in. high, standing position, painted hair and features, jointed arms. 1 doz. in box...........Doz. **78c**

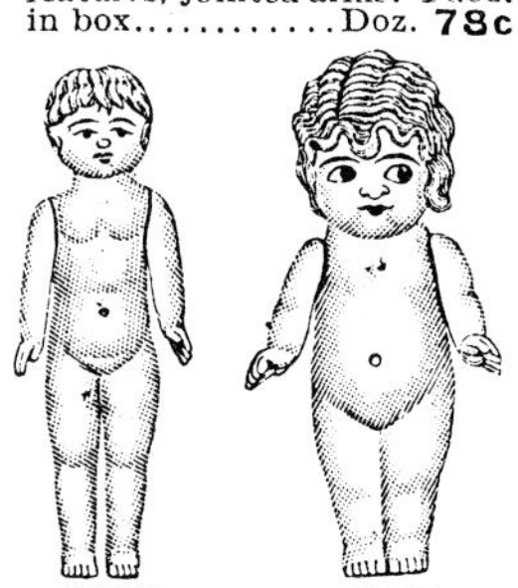

F9528 F9527

F9528—6 in. high, standin position, painted features an hair, jointed arms. 1 doz. i box...........Doz. **87**

F9527—6 in. high, standir position, painted featur and hair, jointed arms. 1 do in box........Doz. **95**

1923

Brightly Colored Celluloid Toys

Will keep baby amused when bathing. Two pretty swans, 2¾ inches long; two ducks, 2¾ inches long and two gold fish, 3⅝ in. long. Fast colors. Shipping weight, 12 ounces.

69F9178—Box of **6**.......... **39c**

Celluloid Teething Toys Colors Will Not Come Off

Cute doll, 4½ in. high, barrel shape red rattle, 2¾ in. diameter with twisted handle, 2⅛-inch teething ring with cord and tassel.
Shipping weight, 8 ounces.
69F9153..................... **29c**

What's Inside Surprise Box?

Unhook the lid, quick as a wink up it comes, with a funny squeak. Wood box covered with paper. Size, 4⅛x4⅛ inches. Shipping wt., 10 oz.

69F7020 **29c**

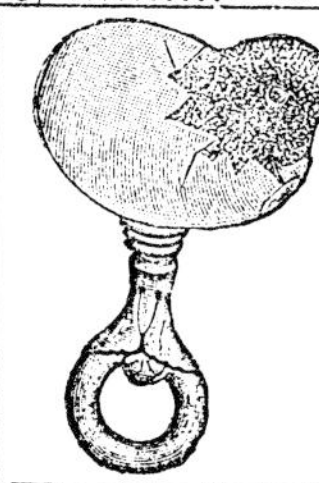

The Ever Popular Rattle

Three in one. Rattle, whistle and teething ring. White rubber. Height, over all, 4¼ inches. Shipping weight, 3 oz.

49F4400. **15c**

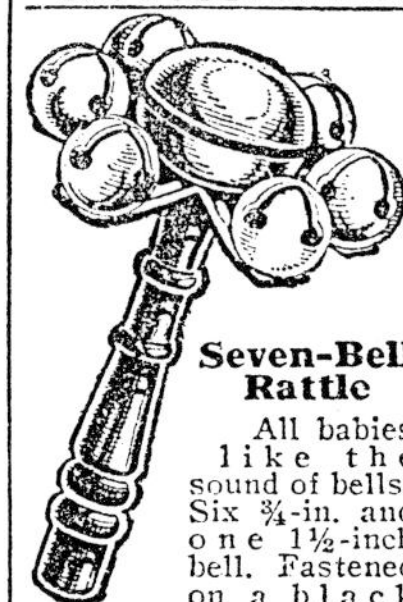

Seven-Bell Rattle

All babies like the sound of bells. Six ¾-in. and one 1½-inch bell. Fastened on a black wood handle. Height, over all, 5 in. Shpg. wt., 8 oz.

49F2480... **12c**

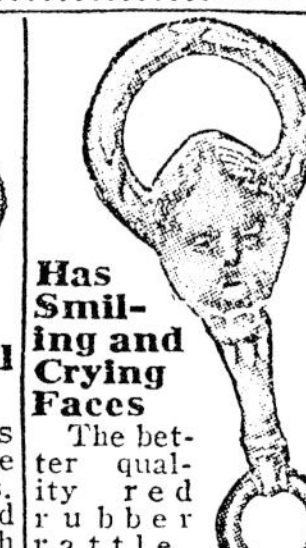

Has Smiling and Crying Faces

The better quality red rubber rattle. Lgth., over all, 7 in. Shpg. wt., 4 oz.

49F4403 **42c**

1926

ZYLONITE RATTLES

1F4518—5¾ in. long, pink & white combination with assorted Mother Goose pictures and rhymes, twisted celluloid handles. Asstd. 1 doz. in box. Doz **89c**

1F4490—3 styles, cerise, 6 in. long, octagon, ball and fancy round shape. Asstd. 1 doz. in box...........Doz **89c**

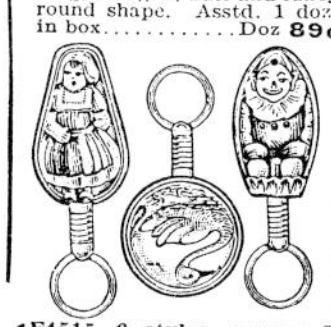

1F4515—6 styles, average 6 in. long, round and egg shapes with embossed and hand painted juvenile characters in assorted pastel colors, flexible neck and teething ring. Asstd. 1 doz. in box........Doz **$2.10**

HAND PAINTED CELLULOID RATTLES

Transparent Eyes and Rolling Pupils

1F4483 1F4481

An exceptionally attractive novelty rattle. Heavy celluloid, pink and blue color combinations with hand painted features in delicate pastel colors. With strong teething ring.

1F4483—6 in. long. Asstd. 1 doz. in box.. Temp. Out
1F4484—8 in. long. Asstd. ½ doz. in box.Doz **$4.00**

BELL RATTLES

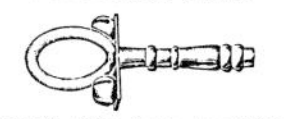

1F2457—6 in. long, two ¾ in. double slot nickeled bells attached to heavy celluloid crosspiece, pink and blue enameled wood handle, fine white rubber teething ring at end. 1 doz. in box. Doz **84c**

1F2458—6 in. long, three 1¼ in. double slot nickeled bells attached to celluloid band, pink and blue enameled wood handle. 1 doz. in box. Doz **87c**

RATTLES WITH VOICES

Two-In-One Toy

1F4520—2 styles, average 4 in. long, assorted juvenile characters in pastel color combinations, concealed voice. When pressed, rattle cries. Asstd. 1 doz. in box. Doz **$2.00**

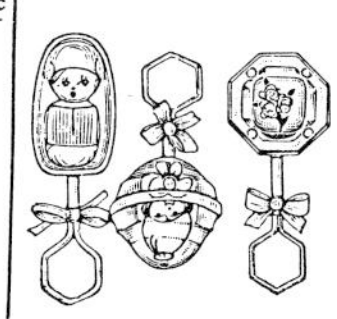

1F4519—2 styles, average 6 in. long, fancy shapes with assorted juvenile characters in pastel color combinations, ribbon bows, concealed voice. When handle is pressed, rattle cries. Asstd. 1 doz. in box...........Doz **$2.00**

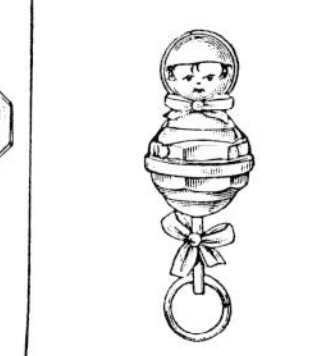

1F4521—2 styles, average 6½ in. long, juvenile characters in pastel colors, ribbon bows, teething rings. When handle is pressed, rattle cries. Asstd. ½ doz. in box. Doz **$3.95**

1925

33 Piece "French Ivory" Infants' Toy Assortment

First quality clean French Ivory stock with pink or blue silk ribbon bows. Feature them to your Holiday trade

Asst. 33 Pieces $8.95

Asst. 33 Pieces $8.95

F4515—33 pieces, asstd. infants' play sets, 18 rattles and 15 asstd. pieces, first quality clean French Ivory stock, with pink or blue silk ribbon bows, blue marbleized white lined boxes, 33 pieces in asst. Each asst. in specially constructed carton to assure arrival in perfect condition.

ASST. (33 pcs.) **$8.95**

BUTLER BROTHERS, RANDOLPH AND CANAL STREETS, CHICAGO

1922

IMPORTED ZYLONITE TOYS.

An extremely popular line of up-to-date novelties. All exceptional values considering **quality** and **finish.**

Rolly-Polies.

Weighted bottoms, always right themselves

2F3141

2F3142

2F3141—2¼ in., 4 styles, alligator, cat and dogs, natural colors. 1 doz. in box, asstd.......Doz. **42c**

2F3142—2¾ in., 12 styles, up-to-date characters, painted features and costumes. 1 doz. in box, asstd............Doz. **72c**

FLOATING ZYLONITE FIGURES.

For baby's bath, aquariums, etc.

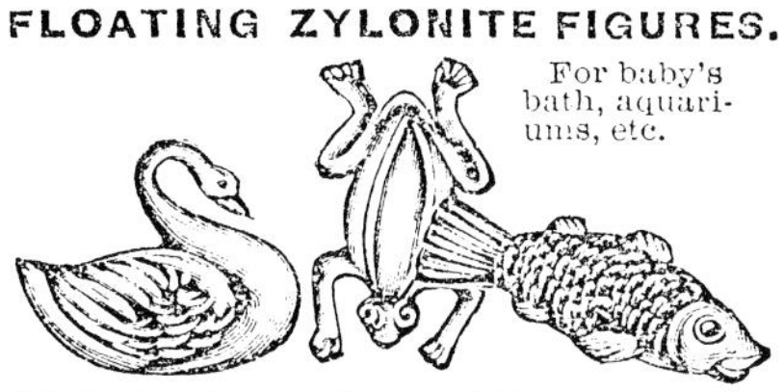

2F3148—About 2¼ in., 4 styles, natural colors, float in lifelike way. 1 doz. in box asstd. Doz. **36c**

2F3149—3 in. 1 doz. in box...........Doz. **79c**

2F3150—4½ in. ½ doz. in box.......Doz. **$1.75**

1911

ZYLONITE DOLLS.

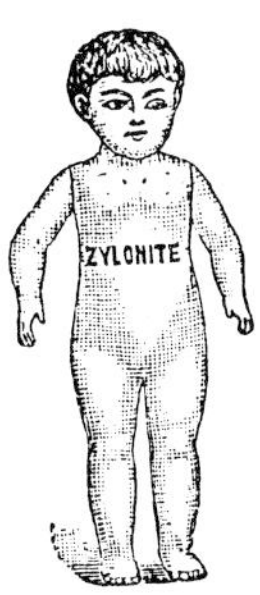

Jointed Arms — Light wt., almost indestructible flesh tinted, hollow, painted hair, eyes and eyebrows, free legs. 1 doz. in box, unless stated.

1F4090—3 in.........Doz. **42c**

1F4091—4⅜ in....... " **78c**

1F4093—7½ in. ½ doz. in box. Doz. **$2.25**

1F4145—8¾ in. 1 in box, ⅓ doz. in pkg............Doz. **$3.60**

1F4094—10 in. ⅓ doz. in box. Doz. **$4.25**

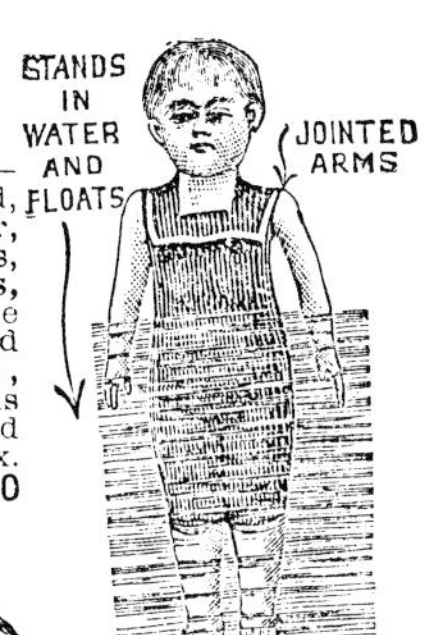

1F4098, Floating—5½ in., flesh tinted, hollow, painted hair, eyes and eyebrows, **jointed arms,** asstd. red and blue painted trunks and bathing suits, weighted feet, stands upright in water and floats. ½ doz. in box. Doz. **$1.90**

Jointed Arms and Legs — *Very unusual values. Splendid sellers.* Painted shoes and socks.

1F4096—8¼ in. 1 in box, ⅓ doz. in pkg. Doz. **$4.25**

1F4097—11 in. 1 in box Doz. **$7.50**

1911

INFANTS' CELLULOID PLAY SETS

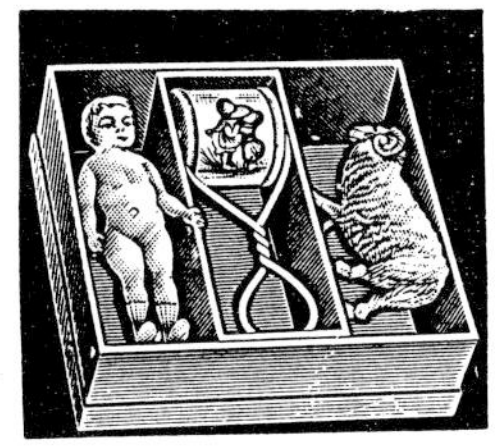

F4477—4 pieces, 5 in. doll, tinted features, jointed arms, 6 in. barrel rattle with twisted handle, 2½ in. high bouncing 2 color ball, 4½ in. sheep, all ribbon tied. ½ doz. sets in pkg......Doz. sets, **$4.50**

F4479—3 pieces, 5 in. doll, tinted features, jointed arms, ribbon bow, 6¼ in. barrel rattle with twisted handle, 4¾ in. sheep rattle. ½ doz. sets in pkg......Doz. sets, **$4.50**

F4478—5 pieces, 5 in. doll, tinted features, jointed arms, 6 in. barrel rattle, with twisted handle, 3½ in. high, bouncing 2 color ball, 4½ in. sheep, 3¾ in. tinted swan, ribbon tied. ½ doz. sets in pkg. Doz. sets, **$9.00**

1922

Zylonite Rattle Assortment

6 styles, heavy celluloid, attractive color combinations

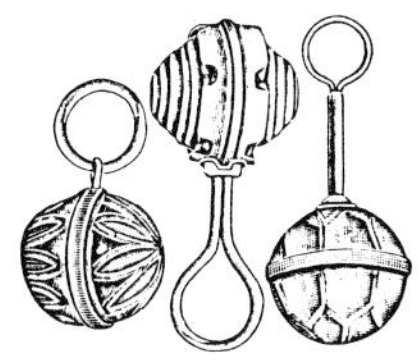

F4463—6 styles, aver. 5½ in., large rattle, ball, oval, octagon and novelty shapes, some with long handle, others with teething ring. Asstd. pink and white, blue and white, cerise and white combinations. 1 doz. in box. Doz. **89c**

Hand Painted Zylonite Rattle Asst.

A carefully selected assortment of the best selling numbers. First-class merchandise that you can retail for 25c at a profit.

F4476—5 styles, aver. 5½ in., egg, barrel, round and fancy novelty shapes, hand decorations and pastel tints, for-get-me-nots, storks and asstd. juvenile designs, large pink or blue bows, solid pink and white and blue and white combinations. Asstd. 1 doz. in partitioned box. Doz. **$2.00**

1922

FLOATING ZYLONITE TOYS

Natural colors, floating, for baby's bath, aquariums, etc.

4503—3 styles, aver. 1¾ in. swan, fish and duck. Asst. 2 doz. in box....Doz. **35c** (Total 70c)

4490 — 5 styles, aver. 2⅜x1¾ in., standing position, sheep, cow, horse, donkey and pig. Asstd. 1 doz. in partitioned box. Doz. **75c**

F4500 — 6 styles, aver. 3½x2½ in., swan, duck, fish, turtle, frog and lobster. Asstd. 1 doz. in box. Doz. **84c**

F4495—5 styles, aver. 3½x2¼ in., standing position, dog, sheep, cow, horse and goat. Asstd. 1 doz. in partitioned box............Doz. **84c**

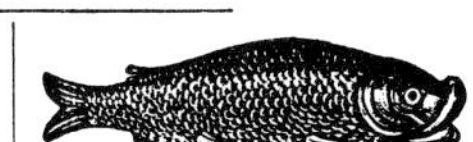

F4505 — 3 styles, aver. 4¼x2¼ in., swan, duck and fish. Asstd. 1 doz. in box. Doz. **$1.25**

F4507 — 3 styles, aver. 5½x3⅛ in., duck, swan and fish. Asstd. ½ doz. in partitioned box. Doz. **$1.98**

510—6 styles, aver. 5x3¾, standing position, dog, sheep, bear, cow, horse and elephant. Asstd. ½ doz. in partitioned box.................................Doz. **$2.15**

FLOATING ZYLONITE BATTLE SHIP

F4506—5¼ in., painted gray good model, gun turrets, latest type masts. ½ doz. in box.............Doz. **95c**

FIGHT FOR THE TURN. Buy of each item a safe minimum. Place orders every day. Let the jobber carry your surplus stocks. **FIGHT FOR THE TURN.**

1922

1932

1932

1932

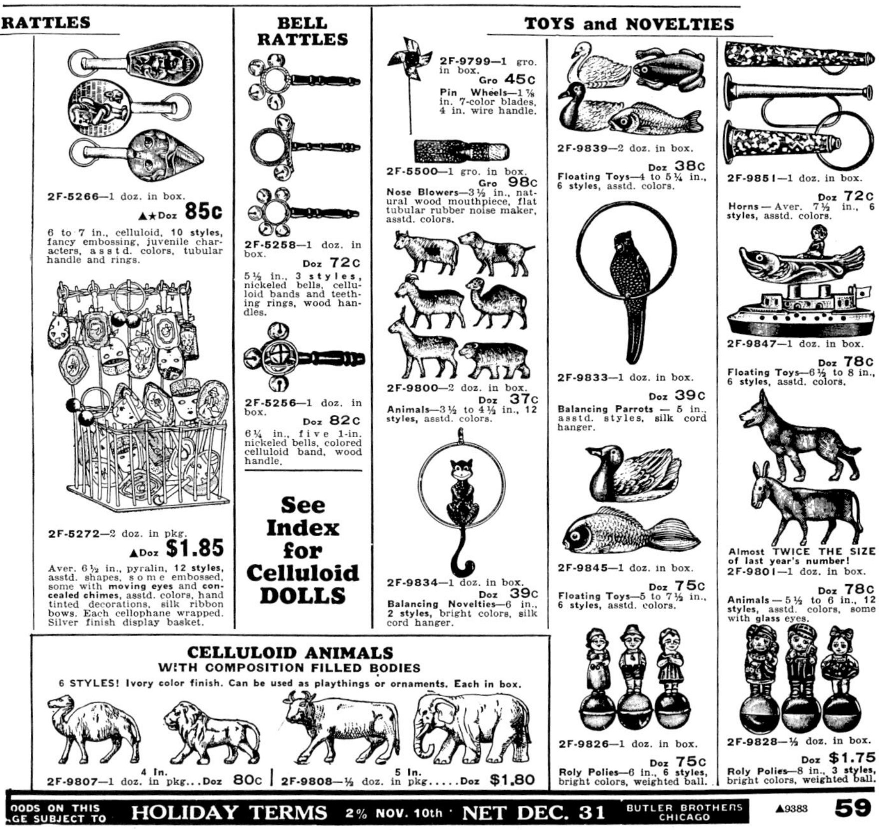

1932

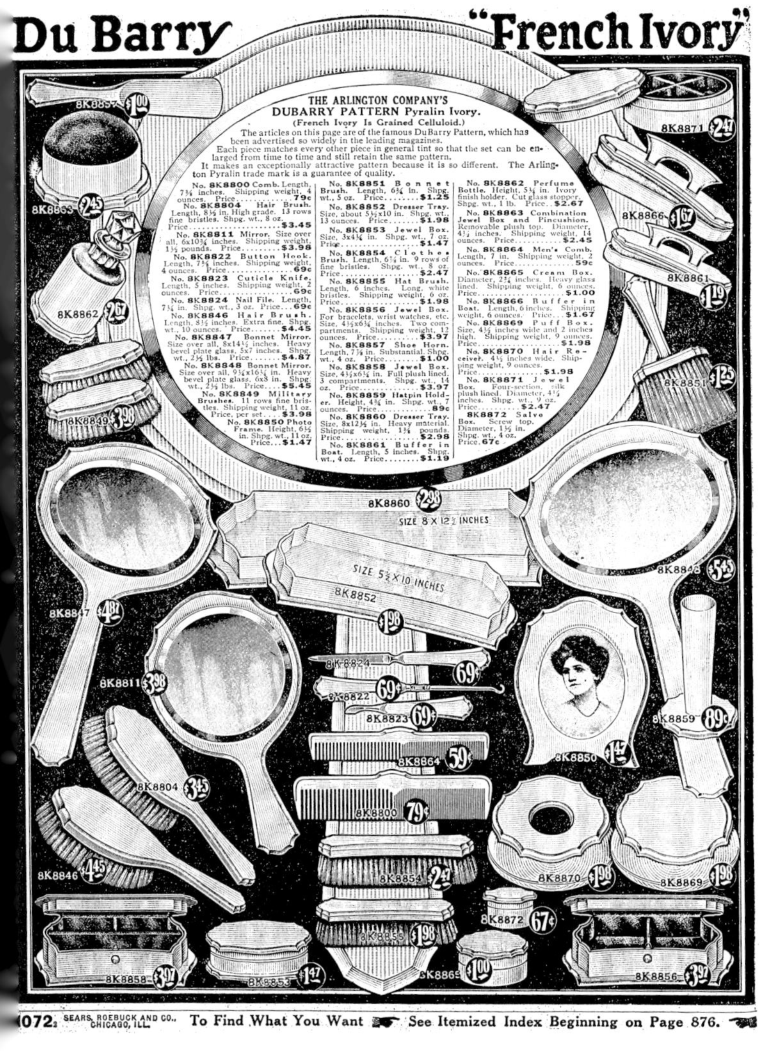

1908

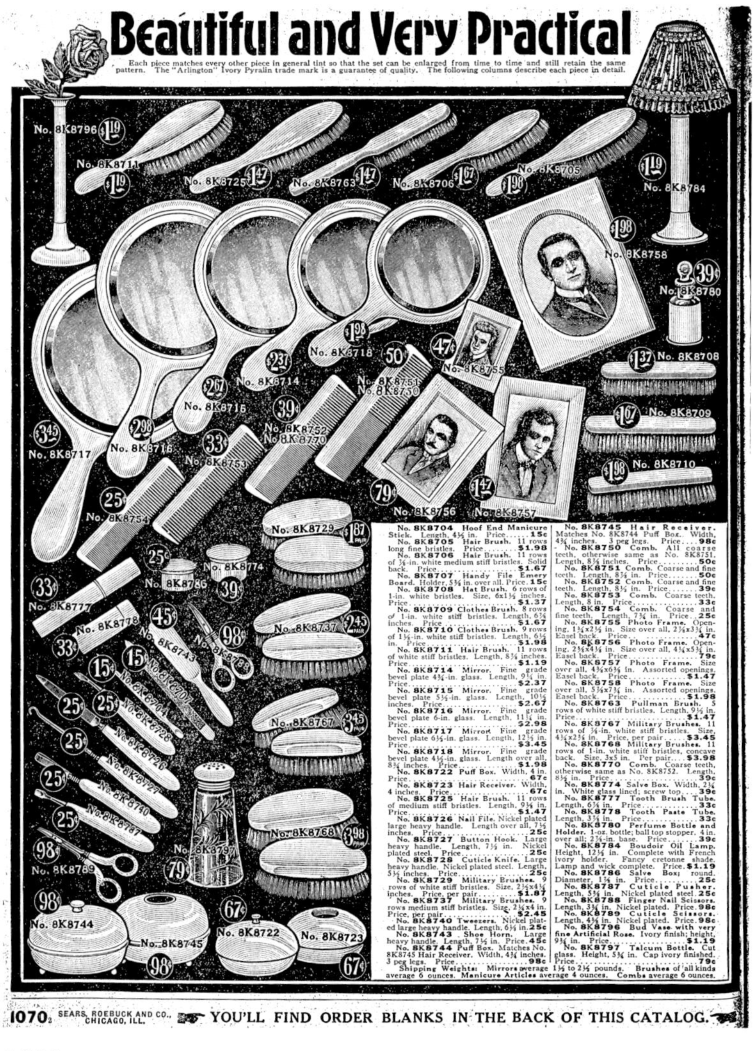

1908

Bibliography

Anderton, Johana Gast. *More Twentieth Century Dolls - from Bisque to Vinyl.* Volume One, A-H. Des Moines, Iowa: Wallace-Homestead Book Co., 1983.

Baekeland, L. H. "The Invention of Celluloid." *Scientific American Supplement.* 28 March 1914: 202-203.

Belanger, Al. "An Interesting Art and Craft Medium." *Industrial Arts & Vocational Education.* September, 1939: 293-6.

Boyer, Jacques. "The Manufacture of Celluloid." *Scientific American.* 30 April 1910.

Buchholz, Shirley. *Century of Collectors Dolls.* Cumberland, Maryland: Hobby House Press, 1983.

"Celluloid: The Oldest Plastic Maker Molds Brighter Future." *Newsweek.* 6 June 1936: 34-6.

Clark, Ronald W. *The Scientific Breakthrough: The Impact of Modern Invention.* New York, NY: G. P. Putnam, *1974.*

DuBois, Harry. *Plastics History U.S.A.* Boston, Massachusetts: Cahners Books, 1972.

Ebner, Professor Doctor. "The Riddle of Cellulose." *The Living Age.* 5 June 1926:513-16.

Dunn, Shirley. *Celluloid Collectibles: Identification and Value Guide.* Paducah, Kentucky: Collector Books, 1996.

Friedel, Robert. *Pioneer Plastic: The Making and Selling of Celluloid.* Madison, Wisconsin: The University of Wisconsin Press, 1983.

Gould, David C. and Donna Crevar-Donaldson. *Toys: Occupied Japan.* Gas City, Indiana: L-W Book Sales, 1993.

Hanlon, Bill. *Plastic Toys: Dimestore Dreams of the '40s and '50s.* Atglen, Pennsylvania: Schiffer Publishing, 1993.

Katz, Sylvia. *Plastics: Common Objects, Classic Designs.* New York: Harry N. Adams, Inc., 1984.

King, Constance Eileen. "The Search for an Indestructible Substance." *The Collector's History of Dolls.* New York: Bonanza Books, 1977.

Kitahari, Taruhisa. *Yesterday's toys: Celluloid Dolls, Clowns, Animals.*

Mc Fadden, Sybill. "Character Celluloids: Collectible Dolls of the Future." *Doll News.* Fall 1976: 27-30.

Lindenberger, Jan. *Collecting Plastic: A Handbook & Price Guide.* Atglen, Pennsylvania: Schiffer Publishing, 1991.

McNultly, Lindi Stewart. *Price Guide to Plastic Collectibles.* Radnor, Pennsylvania: Wallace Homestead, 1987.

Meikle, Jeffrey L. *American Plastic: A Cultural History.* New Brunswick, Canada: Rutgers University Press, 1997.

DiNoto, Andrea. *Art Plastic-Designed for Living.* New York: Abbeville Press, 1984.

Reily, Julie A. "Celluloid Objects: Their Chemistry and Preservation." *Journal for the American Institute for Conservation.* 30 (1991): 145-162.

Robinson, Julie. "John Wesley Hyatt and the invention of Celluloid." *Antique Week Central Edition.* 19 June 1995: 1-2+.